PENGUIN HANDBOOKS

THE TRAVELER'S GUIDE
TO THE VINEYARDS
OF NORTH AMERICA

Writer, lecturer, and photographer, William I. Kaufman is the creator of over a hundred books on wine, cooking, travel, and entertaining. His most recent book, *The Whole-World Wine Catalog*, is published by Penguin Books. Some of his other works are *Champagne, Perfume, The Wonderful World of Cooking* (four volumes), and *Cooking in a Castle*. His books have been reprinted in many editions and translated into numerous languages. Mr. Kaufman is the recipient of many foreign honors, including being made a Chevalier de l'Ordre du Mérite Agricole by the French Republic. One-man exhibitions of Mr. Kaufman's photographs have been seen in every major city of the United States. He is often referred to as The Idea Factory.

The Traveler's Guide to the Vineyards of North America

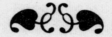

by William I. Kaufman

PENGUIN BOOKS

Penguin Books Ltd, Harmondsworth,
Middlesex, England
Penguin Books, 625 Madison Avenue,
New York, New York 10022, U.S.A.
Penguin Books Australia Ltd, Ringwood,
Victoria, Australia
Penguin Books Canada Limited, 2801 John Street,
Markham, Ontario, Canada L3R 1B4
Penguin Books (N.Z.) Ltd, 182–190 Wairau Road,
Auckland 10, New Zealand

First published 1980

Copyright © William I. Kaufman, 1980
All rights reserved

LIBRARY OF CONGRESS CATALOGING IN PUBLICATION DATA
Kaufman, William Irving, 1922–
The traveler's guide to the vineyards of
North America.
(Penguin handbooks)
1. Wine and wine making—United States—Directories.   2. Wine and wine making—Canada—Directories.   3. Wine and wine making—Mexico—Directories.   I. Title.
TP557.K38   917.3'04'926   79–21004
ISBN 0 14 046.399 2

Printed in the United States of America by
Offset Paperback Mfrs., Inc., Dallas, Pennsylvania
Set in CRT Helvetica and Helvetica light

Except in the United States of America,
this book is sold subject to the condition
that it shall not, by way of trade or otherwise,
be lent, re-sold, hired out, or otherwise circulated
without the publisher's prior consent in any form of
binding or cover other than that in which it is
published and without a similar condition
including this condition being imposed
on the subsequent purchaser

To Rose,
her golf clubs,
her tennis racket,
and her caviar

Contents

Introduction	7
California: The Southern Region and Baja California	13
California: The Central Coast and the San Joaquin Valley	19
California: The East Bay Region and the Livermore Valley	39
California: The Napa Valley	53
California: The Sonoma Region	69
The Pacific Northwest (Oregon, Washington, Idaho, and British Columbia)	89
The Ohio River Valley (Ohio, Indiana, and Kentucky)	99
The Great Lakes Region (Minnesota, Wisconsin, Michigan, Illinois, Indiana, Ohio, Pennsylvania, New York, and Ontario)	107
New York: The Finger Lakes District	127
The New England States (Connecticut, Massachusetts, Rhode Island, and New Hampshire)	133
New York: The Hudson Valley	139
The Mid-Atlantic Region (Pennsylvania, New Jersey, and Maryland)	145

The Southern States (Virginia, North Carolina, and South Carolina)	155
The Upper Mississippi Region (Iowa and Missouri)	161
Addendum	169
Index of Vineyards	193

Introduction

It is true—fabulous wines *are* made in California and New York! The problem is that it is difficult to know which the great ones are, or if they are even available. One pays for a French Red wine that tastes a bit sour, but could the cheaper American vintage listed on the menu *really* have been any good? If such questions are beginning to interest you, or frustrate you, it may be time to learn more about American wines, which are only now fully recovering from a great man-made disaster of some decades ago.

From 1920 until 1933 the Eighteenth Amendment stipulated that no alcoholic drink, including, of course, wine, could be produced or sold for popular consumption in the United States. Until then wine had played as prominent a role in American domestic and festive habits as it does today. It was a subject of literature and of intellectual curiosity, and while it was never a common staple, wine was recognized as one of the pleasures money could buy. Of course, the vintners of France, Germany, and Japan didn't stop making wine during Prohibition, nor did people all over the world stop drinking it. At Repeal, foreign products, with no recent history of interruption in the growing and aging process, were available to Americans again. Numbers of Americans drank European wines, but there was no vast volume in wine sales after 1920. It was around this time, and until after World War II, that a few individuals and families remembered that many American regions produced fine wines once, and they set out, in many cases from scratch, to do so again.

Since the 1850s wine had been made in the United States on a respectable scale. Efforts at wine making in the seventeenth and eighteenth centuries had failed because European grape stock could not withstand North American winters and pests. When vintners began cultivating native-American grapes, however, production of fine wines on a large scale became possible here. By the middle of the last century a second element was added to the industry—the participation of immigrants in American wine making. People of Italian, French, Swiss, German, and other European origins were making wines the way they had in Europe, with grapes that would grow here. Their knowledge added much, and their family wineries, or the vineyards where they worked, began producing wines that by 1900 won prizes all over the world and enjoyed a strong European reputation. It was this tradition and this continuum of experience that Prohibition interrupted.

Various factors have converged in the past twenty years to create a revival of popular interest in wines. It is almost as if the wine idea were now spreading from west to east, since the heart of the renewed activity has been in California, with much competition from strong wine states like New York, and from former centers of the last century's viticulture, like Ohio. Growers began growing again, going back to vineyards they had held onto since before Prohibition, or, like many in California, buying disused vineyards and revivifying them. They began producing vintages that, like their great forebears, were of a quality to challenge European products. This surprised and excited American wine lovers who knew nothing about the distinguished nineteenth-century history of American wines. America began to remember its own wine history precisely when leisure spending by the young was beginning to boom.

Three categories of growers can be identified among those responding to inquiries for this book. First are the family-owned operations, often founded by an ancestor before 1900. Succeeding generations of the family worked the land more or less continuously, and during Prohibition at least maintained its wine-producing potential. A pride and continuity of achievement survives at many of these vineyards, which are enjoying new prosperity in the context of a long tradition. The second category is comprised of giant combines and large, mass-market–oriented wine companies, such as Almadén, Taylor, and Inglenook, to name but a few. The great scale of these operations does not preclude the possibility of producing fine vintages along with the decent, popularly priced ones. The third category of growers would be individuals, families, and other groups of people, often young, who have chosen to live on the land and to make a living or keep busy by making wines—and it is to individuals like this that we owe the revival of many fine abandoned vineyards after Prohibition.

The environment that surrounds wine growing on a large scale, a community scale, or an isolated scale, is generally a pleasant one, involving as it does a mildness of climate on hospitable farmland. Wineries are wonderful places to visit. Californians and New Yorkers have discovered it; the French have long known that wine making encourages a very pleasant kind of touring. Many wine makers and winegrowers who contributed to the entries in this book are conscious of the importance of fostering this kind of interest in their métier. These people tell of the long history of wine making in their parts of the country. Of course these include Californians and New Yorkers, but growers from Ohio, Massachusetts, Missouri, Kentucky, Virginia, and other states are also represented.

The book is ordered so that car trips of a day or more can be undertaken to groups of vineyards. The listings of vineyards are organized to encourage travel among neighboring counties, or even states, which share a common grape-growing environment with identifiable characteristics. Division into districts seems most often to be made according to the proximity of a body of water, or to the contours of a valley or a range of hills. California's northern valleys, Sonoma, Napa, and Livermore, share sea moisture, hot days, and bright sun. All three valley areas are accessible from San Francisco in journeys of a day or more. The Ohio River Valley provides a fairly uniform environment for vineyards in three states, and it is important to understand the wine districts along the Ohio in terms of the climates and history they share. New England wineries that told us they like to entertain are widely spaced in a way conducive to one's taking several weekend tours. Touring the Northwest districts is enhanced by the fact that growing is done in several concentrated areas. The Mission Country districts of the California coast are singularly well favored by nature and history to entertain travelers passing down the coast. Visitors may wish to turn again and drive north through the districts of the San Joaquin Valley. New York State's wine districts fall into two distinct geographical areas which should be visited separately. In any entry about wine-growing districts in Florida, Michigan, or any other state, the practice of those vineyards which encourage curiosity has been to tell readers where they can stay overnight, where there is a good meal, and what contributes to the most pleasant picnicking and traveling in the region.

Each entry gives the complete address of the vineyard or winery, and the name of the person to contact should an appointment be necessary. The vineyard's hours and the schedule of tours, where applicable, are listed, as well as the following:

> **WT**—Wine tasting available
> **RS**—Retail sales available
> **P**—Picnicking encouraged
> **D**—Dining available at the vineyard
> **WP**—Wines produced
> **GR**—Grapes and/or wines
> **R**—Restaurants
> **HS**—Historical sites nearby
> **HM**—Hotels and/or motels in the area

In the case of vineyards, we have listed the grapes grown; in the case of wineries, a list of wines produced is offered. Finally, we have presented a capsule history with each entry to give a picture of what the visitor might anticipate.

It is inevitable that the occasional restaurant will be awful, the motel room notable for its lack of closets; conversely, you might be treated to a view of a sunrise that you will never forget and a healthy country breakfast in bed. Picnics should be planned around the local cuisine and the wines that you are going to encounter, and should enhance your new knowledge about wine making, wine growing, and the tastes of American wines. You may pick up wine-making equipment for yourself, and grapes with which to do it, and find yourself in the pages of a book like this someday in the future. We hope that this will be one of the books you regularly consult when planning any sort of trip around the country, for wines can provide a very special introduction to the regions in which they grow.

The Traveler's Guide to the Vineyards of North America

California: The Southern Region and Baja California

The California wine industry was born in Southern California. In 1769 the famous missionary Fra Junípero Serra planted the first vineyards in the state at Mission San Diego. The fathers produced wine for their own use, and also sold small quantities to American and Spanish settlers in the area. Among the first large growers to follow the fathers' lead were Jean Louis Vignes and Pierre Sainsevain, who imported vines from Europe and grew them so successfully that within a few years they were shipping wines to the East and abroad. The growth of the city of Los Angeles has displaced many of the older vineyards; a large vineyard once flourished, for instance, on the present site of Los Angeles's Union Station. Winegrowers once centered their efforts mainly in the Cucamonga Valley, forty-five miles east of Los Angeles, among the towns of Cucamonga, Etiwanda, Ontario, and Guasti. This rich wine region has a great tradition that dates back to the growing done by Secondo Guasti in the early 1900s. He brought whole families from Italy to help him transform the Cucamonga desert into rich wine land, and his Italian Vineyard Company prospered. By the 1940s the district's wineries could boast several distinctive local wines, unique because of the singular growing climate in the area.

As big industry and urban smog have slowly ruined this district for large-scale growing of the traditional kind, vintners have moved principally to two areas: around Escondido, near San Diego, and to lands surrounding Temecula in southwestern Riverside County. The vineyards, either on old lands or new, share a unique grape climate and produce a similarly unique selection of wines. Both large-scale and smaller growers thrive in Southern California, and you can arrange several day-trips to include both kinds of operations.

Old Guasti, in the Cucamonga Valley, has a vivid and well-maintained past as a wine center, and offers a survey of the region's wine history as well as a good example of the wine making as it goes on today. But trips a bit farther afield may be the most interesting in Southern California for the wine enthusiast, since you can see for yourself the transformation of a great old wine district as it retrenches, and in some cases, relocates, in the face of the voracious urban sprawl nearby. The wines that have been produced in this region, and that continue to be produced, have always been worth the wine lover's attention.

BERNARDO WINERY / 13330 Paseo del Verano Norte, San Diego, California 92128　　　　　　　　　　Ross Rizzo: (714) 847-1866
OPEN DAILY 7 A.M.–6 P.M. WT, RS, P, D.

GR Zinfandel; Muscat; Tokay; Carignane; Mission; Black Morocco; Golden Chasselas　**R** The Other Place (at the winery); Rancho Bernardo Inn　**HM** Rancho Bernardo Inn; TraveLodge; Mt. Vernon Motel　**HS** Palomar Observatory; San Diego Wild Animal Park.

MISC Founded in 1889, the Bernardo Winery is now owned by Ross Rizzo. This is the last old producing winery and vineyard in San Diego County. The winery features gift and craft shops open Wednesday–Sunday 11 A.M.–5 P.M.

CALLAWAY VINEYARD AND WINERY / 32720 Rancho California Road, Temecula, California 92390　　　　　　　　　　(714) 676-4001
OPEN TUESDAY–SUNDAY NOON–4 P.M.; CLOSED MAJOR HOLIDAYS. RS, P.

GR Chardonnay; Chenin Blanc; Sauvignon Blanc; White Riesling; Cabernet Sauvignon; Zinfandel; Petite Sirah　**R** Rancho California Inn; Rancho California Golf Resort　**HM** Rancho California Inn; Rancho California Golf Resort　**HS** Temecula is the home of the Temeku Indian tribe and an early stagecoach station.

CALIFORNIA: THE SOUTHERN REGION AND BAJA CALIFORNIA 15

MISC The Callaway Vineyard and Winery produced its initial vintage in 1974. All vines are planted on *Vitis vinifera* roots.

FERRARA WINERY / 1120 West Fifteenth Avenue, Escondido, California 92025
Vera M. Ferrara: (714) 745-7632

OPEN MONDAY–FRIDAY 9 A.M.–6:30 P.M.; SATURDAY AND SUNDAY 10 A.M.–6:30 P.M.; CLOSED MAJOR HOLIDAYS. APPOINTMENTS FOR GROUPS OF TEN OR MORE. WT, RS.

WP Dry and Sweet wines, generic and varietal   **R** Fireside Restaurant; Marie Calenders   **HM** Mt. Vernon Motel; Laurel Inn   **HS** San Luis Rey Mission; Pala Mission; San Pasqual Battlefield Monument.

MISC The Ferrara Family, now in its third generation of wine making, has been in Escondido since 1919. The winery has been designated by the State of California as Escondido's historical point of interest.

J. FILIPPI VINTAGE CO. / P.O. Box 2, Mira Loma, California 91752
(714) 984-4514

OPEN DAILY 9 A.M.–6 P.M.; CLOSED NEW YEAR'S DAY, THANKSGIVING, CHRISTMAS. WT, RS, P.

GR Grenache; Mataro; Zinfandel; Golden Chasselas; Burger; Carignane; Mission   **R** Sycamore Inn; Cask and Cleaver; Arbor Restaurant   **HM** Holiday Inn   **HS** Casa de Rancho Museum; Thomas Vineyards; Historical landmarks.

MISC The Filippi Family has been producing vintners for generations. It was not until 1922, however, that Giovanni Filippi and his son Joseph left their native northern Italy to come to Cucamonga Valley. By 1934 the first portion of the present winery was built, and from a small beginning, the winery has increased to an annual capacity of over three million bottles of wine. Through the Filippi idea of "winery direct to consumer," retail sales and tasting rooms have been established throughout Southern California. They are in San Gabriel Valley: 9613 Valley Boulevard, El Monte; in Eastern Los Angeles: 8447 Rosemead Boulevard, Pico Rivera; in Southwest Los Angeles: 5107 West El Segundo Boulevard, Hawthorne; in North San Diego County: 840 East Vista Way, Vista; in San Fernando Valley: 8253 Sunland Boulevard, Sun Valley; and in Orange County: 12872 South Harbor Boulevard, Garden Grove.

GALLEANO WINERY / 4231 Wineville Road, Mira Loma, California 91752
Don Galleano: (714) 685-5376

OPEN MONDAY–SATURDAY 8 A.M.–6 P.M.; CLOSED MAJOR HOLIDAYS. WT, RS.

Galleano Winery (cont.)

GR Zinfandel; Mataro; Grenache; Carignane; Muscat; Golden Chasselas; French Colombard; Barbera   **R** Sycamore Inn; Magic Lamp.

MISC Three generations of the Galleano Family are currently active at this winery. The Galleano Ranch is situated on lands purchased from General Cantu, former governor of Baja California.

MARTIN WINERY, Inc. / 11800 West Jefferson Boulevard, Culver City, California 90230   (213) 390-5736

OPEN DAILY 11 A.M.–7 P.M.; CLOSED NEW YEAR'S DAY, EASTER, CHRISTMAS. WT, RS, D.

WP 180 wine types are available in the tasting room and café
R Cyrano's; Gulliver's; Gatsby's; Hungry Tiger; Charley Brown's
HM Howard Johnson's Airport Lodge; Ramada Inn.

MOUNT PALOMAR WINERY / 33820 Rancho California Road, Temecula, California 92390   John Poole or Joe Cherpin: (714) 676-5047

OPEN DAILY 9 A.M.–5 P.M.; CLOSED NEW YEAR'S DAY, THANKSGIVING, CHRISTMAS. WT, RS, P.

GR Cabernet Sauvignon; Petite Sirah; Shiraz; Zinfandel; Chenin Blanc; Sauvignon Blanc; White Riesling; Palomino   **R** Homestead Restaurant; Swing Inn Cafe; Scarcella's Pizzeria   **HM** Rancho California Inn; Rancho California Golf Resort   **HS** Temecula is a historical frontier town featuring the Butterfield Stage Stop, a Frontiersman monument, and an Indian reservation.

MISC After twenty years in broadcasting John Poole sold his stations to purchase land, and founded the vineyards in 1969. The winery was started in 1975 with a production of 2,400 gallons, and it has been expanding ever since. In 1976 wine maker Joseph L. Cherpin joined Palomar. Production is now approximately 35,000 gallons per year, from 165 acres under vine. There is a gourmet food shop on the premises.

RANCHO DE PHILO / 10050 Wilson Avenue, Alta Loma, California 91701
Philo Biane: (714) 947-4208

OPEN DAILY BY APPOINTMENT. WT, RS.

GR Mission   **R** Sycamore Inn   **HM** Uplander Motel.

MISC This small winery is run more for pleasure than for profit, and limits annual production to two hundred cases.

CALIFORNIA: THE SOUTHERN REGION AND BAJA CALIFORNIA 17

SAN PASQUAL VINEYARDS / 13455 San Pasqual Road, San Diego, California 92025 David J. Allen: (714) 741-0855

OPEN 9 A.M.–4 P.M. BY APPOINTMENT ONLY; CLOSED MAJOR HOLIDAYS. RS.

GR Chenin Blanc; Sauvignon Blanc; Muscat Blanc; Petite Sirah; Gamay **R** Cask and Cleaver **HM** Escondito Motor Hotel **HS** Site of Battle of San Pasqual.

MISC Vineyards planted in 1974; winery constructed in 1976. First release, 1978.

SOUTH COAST CELLARS / 12901-B South Budlong Avenue, Gardena, California 90247 Doug Anderson: (213) 324-8006

OPEN WEEKENDS AND EVENINGS BY APPOINTMENT. WT, RS.

WP Zinfandel; Cabernet Sauvignon; Merlot.

MISC Owner and wine maker Doug Anderson writes: "South Coast Cellar was established in 1977 to explore the interface of 'old-style' Bordeaux winemaking practices with red varietals grown along the southern and central California coast. Consumers interested in big, young reds are invited to become part of our history. First releases in 1979."

THOMAS VINEYARDS / 8916 Foothill Boulevard, Cucamonga, California 91730 (714) 987-1612

OPEN MONDAY–SATURDAY 8 A.M.–6 P.M.; CLOSED NEW YEAR'S DAY, THANKSGIVING, CHRISTMAS. SELF-CONDUCTED TOURS. WT, RS, P.

GR Grenache; Salvadore; Zinfandel; Mission; Carignane **R** Sycamore Inn; Magic Lamp **HM** Holiday Inn; Uplander Motel **HS** Rains House; County Historical Museum.

MISC This winery has been established as California's oldest winery, and it is Historical Landmark Number 490. Originally built in 1839 by Tirburcio Tapia on a Mexican land grant, the winery has had many owners and a colorful history. It was purchased by the wine-making Filippi Family and features a distinctive gift shop.

BODEGO DE RANCHO VIEJO, S.A. / P.O. Box 115, 22 Carretera Tecate, Encinada, Tecate, Baja California, Mexico Marcial Ibarra: (903) 399-1560

OPEN MONDAY–FRIDAY 8 A.M.–5 P.M.; CLOSED OFFICIAL HOLIDAYS. WT, P.

Bodego De Rancho Viejo, S.A. (cont.)

GR Rosa del Peru; Mission; Carignane **R** El Dorado **HM** El Dorado Hotel.

MISC This vineyard was founded in 1948.

BODEGAS DE SANTO TOMAS, S.A. / Avenida Miramar 666, Ensenada, Baja California, Mexico Ing. Octavio E. Jimenez: 8-25-09

OPEN MONDAY–FRIDAY; CLOSED OFFICIAL HOLIDAYS. TOURS BY APPOINTMENT. WT, P, D.

GR Mission; Ruby Red; French Colombard; Ugni-Blanc; Chenin Blanc; Pinot Chardonnay; Pinot Noir; Cabernet Sauvignon; Valdepeñas; Carignane; Barbera; Grenache; Rosa del Peru; Palomino; Malaga; Moscato de Alejandria **R** El Rey Sol **HM** Hotel San Nicolas **HS** Ruins of historical missions; La Bufadora.

MISC Founded as a Dominican mission in 1765, this winery was established almost one hundred years ago in 1888.

VIDES DE GUADALUPE (DOMEQ) / Carretera A Tecate, Km. 73, El Sausal, Valle de Guadalupe, Baja California, Mexico
Luis Cetto: 516-44 or 530-31

OPEN MONDAY–FRIDAY; CLOSED OFFICIAL HOLIDAYS. TOURS BY APPOINTMENT. WT, P.

GR Red, Rosé, and White wines.

MISC This vineyard was founded in 1971.

California: The Central Coast and the San Joaquin Valley

After visiting the several fine wineries around San Jose one can easily spend a pleasant day between Santa Cruz, on the coast, and Gilroy, on Highway 101, a few miles inland. At San Jose are the big Masson vineyards, and around Gilroy are a host of fine small vineyards. A bit farther south is Hollister, the home of Almadén Wines. Masson offers tours and tasting; Almadén has two large tasting rooms. This is called Mission Country, all the way down the coast through San Luis Obispo to Santa Ynez. Route 1, the coastline route, takes you south from historic Monterey through Carmel, a lovely, expensive, and tranquil town. The first vintner of the district, Fra Junípero Serra, is buried at the Mission in Carmel where he tended vineyards until his death in 1784. If you travel on Highway 101 you will miss this, as you will be heading directly south to the Paso Robles and Shandon vineyards. To reach Shandon from Carmel, join Route 46 after San Simeon. Around Shandon are some classic central coast vineyards. From these, 101 will take you on through the burgeoning San Luis Obispo wine district, past vineyards in the vicinity of Santa Maria, and into the Santa Ynez district.

The more northerly of vineyard districts along the coast benefit from the cool, moist climate that the Pacific provides; the more southerly districts near Santa Maria and Santa Ynez have less temperate weather. You will see older and more established vineyards in the north, and more experimentation and innovation on smaller, newer vineyards east of the San Luis Obispo area. Farther south, below Santa Maria, you can observe growing operations that have had to move there, relocating from the smog-infested suburbs of Los Angeles. You will find both historic and scenic attractions all along the way (although Highway 101 is an unpleasant experience to be endured where it can't be avoided). Route 1, winding along the coast in the north, provides some of the most beautiful touring in America.

The San Joaquin Valley is also accessible from San Francisco (as well as from the Santa Ynez area, should you be going north again). This "valley" is a two-hundred-mile-long tableland extending south from the Lodi region and intersected by Interstate 5. It is all irrigated land, and here almost a half million acres of grapes produce all the raisins consumed in the United States and half the wine produced in California. It is a mass-production center for wine and grapes and is very interesting as such. The vineyards in this area that do operate on a smaller scale are in the middle of the greatest concentration of grape-growing land in North America.

AHLGREN VINEYARDS / P.O. Box 931, Boulder Creek, California 95006
Dexter Ahlgren: (408) 338-6071

WINERY VISITS BY APPOINTMENT ONLY. RS BY MAIL ORDER.

GR Chardonnay   **WP** Cabernet Sauvignon; Zinfandel; Chardonnay
R The Wild Goose; Buffalo Gals.

MISC This is a small, personal family winery and vineyard that was bonded in 1976.

ALMADÉN VINEYARDS / 1530 Blossom Hill Road, San Jose, California 95118
Tasting Room: Junction of California Routes 152 and 156, Pacheco Pass.
(408) 269-1312

OPEN MONDAY–FRIDAY 10 A.M.–3 P.M. WT.

WP Some fifty-nine different wines including Vintage varietals, Champagne, Sparkling, Dessert, Generic, Brandies, Sherries, Ports, and Vermouths.

CALIFORNIA: THE CENTRAL COAST AND THE SAN JOAQUIN VALLEY 21

MISC The vineyards of Almadén were established in 1852 when Charles Lefranc carefully planted grape cuttings from his native France in the soil near the old Guadalupe Creek, where Almadén's home winery now stands in San Jose. From that modest beginning, marking one of the first successful commercial plantings of fine European wine grapes in Northern California, Almadén today has a total of five wineries and thousands of acres of lush vineyards in a three-county area involving some of the world's most exciting wine districts. Almadén's classic varietals come mainly from the winery's vineyards in Alameda, Monterey, and San Benito counties. Almadén's cooperage among its five wineries exceeds a total capacity of more than 38 million gallons.

Guided tours of Almadén's home winery in San Jose are available to the public every hour on the hour from 10 A.M. to 3 P.M. on weekdays only. The tour spans the old to the new, ranging from Almadén's original winery and huge "Centennial Barrel" (which was displayed both at America's centennial and bicentennial) to one of the most modern bottling houses in the world. Almadén's tasting garden, Don Pacheco, is located at the junction of California routes 152 and 156 near Hollister. Don Pacheco is open to the public from 10 A.M. to 5 P.M. daily, including weekends.

BARENGO VINEYARDS / P.O. Box C, 3125 East Orange Street,
Acampo, California 95220   Michael R. McKee: (209) 369-2746

OPEN DAILY 9 A.M.–5 P.M.; CLOSED NEW YEAR'S DAY, EASTER, THANKSGIVING, CHRISTMAS. WT, RS, P.

GR French Colombard; Chenin Blanc; Ruby Cabernet; Cabernet Sauvignon.

MISC Barengo Winery's heritage dates back to 1868. The vineyards and winery are now owned and operated by Ira and Kent Kirkorian.

BARGETTO'S SANTA CRUZ WINERY / 3535-A North Main Street,
Soquel, California 95073   Patti Ballard: (408) 475-2258

OPEN DAILY 9 A.M.–6 P.M.; CLOSED THANKSGIVING, CHRISTMAS. WT, RS.

WP Premium Dry and Sweet Table wines, and ten Fruit wines
R Shadowbrook; The Courtyard   **HM** Dream Inn; Holiday Inn.

MISC Founded in 1933 by brothers Philip and John Bargetto, second-generation Piedmontese vintners, the winery began selling barrels of wine made in the Italian family tradition, and soon thereafter started producing Premium Red and White Table wines. The winery has won dozens of awards in statewide competitions, and features, among other wines, an extensive list of Fruit wines.

THE BERTERO WINERY / 3920 Hecker Pass Highway, Gilroy, California 95020   Angelo J. Bertero, Jr.: (408) 842-3032

OPEN DAILY 9 A.M.–5 P.M.; CLOSED NEW YEAR'S DAY, EASTER, THANKSGIVING, CHRISTMAS. WT, RS, P.

GR Cabernet Sauvignon; Carignane; Grenache; Barbera; Zinfandel; French Colombard; Pinot Noir   **R** Hecker Pass Inn; Harvest Time Restaurant; Digger Dan's; Mt. Madonna Inn   **HM** Oak Motel; Wagon Wheel; Motel 6   **HS** Gilroy Local Museum; Mount Madonna; two of the largest garlic and onion processing plants in the U.S.

MISC Passed down through three generations, the Bertero Winery still produces wines from the original vineyards, at least seventy-five years old. At the vineyards one can see the nation's oldest and largest live oak tree, estimated at four hundred years of age.

DAVID BRUCE WINERY / 21439 Bear Creek Road, Los Gatos, California 95030   David Bruce: (408) 354-4214

OPEN TUESDAY–FRIDAY BY APPOINTMENT ONLY; SATURDAY 11 A.M.–2 P.M. WT, RS, P.

GR Cabernet; Chardonnay; Pinot Noir; White Riesling   **R** Le Mouton Noir; Plumed Horse; Mountain Charlie's   **HM** Los Gatos Inn

MISC David Bruce, vintner and physician, established his winery in 1964 with homemade equipment and twenty-five acres of vines. The winery now has a fifteen-thousand-gallon capacity.

CALERA / 11300 Cienega Road, Hollister, California 95023
      Josh Jensen: (408) 637-9170 or 637-2344

OPEN SATURDAYS 11 A.M. BY APPOINTMENT. RS, P.

GR Pinot Noir   **WP** Zinfandel; Pinot Noir.

MISC The young Pinot Noir vineyard is located at an elevation of 2,200 feet in the Gavilan Mountains on a limestone outcrop. The winery, closer to town, is a completely gravity-flow winery, with five levels separated by a total of fifty-four feet. The first crush at this new winery was in 1977. Current sales are of Zinfandel, made from grapes purchased from selected vineyards. Pinot Noir sales will begin in 1980.

CALIFORNIA: THE CENTRAL COAST AND THE SAN JOAQUIN VALLEY

CALIFORNIA CONCENTRATE COMPANY / 18678 North Highway 99, Acampo, California 95220   Dennis Alexander: (209) 334-9112

OPEN MONDAY–FRIDAY 8 A.M.–5 P.M. WT, RS.

GR Twenty-five varieties of concentrates including Grey Riesling, Barbera, and Cabernet Sauvignon   **R** Cosmopolitan; Golden Acorn   **HM** Royal Host Inn.

MISC The California Concentrate Company has been family owned since 1935.

CALIFORNIA WINE ASSOCIATION, A. Perelli-Minetti & Sons / P.O. Box 818, Delano, California 93215
           Georganne Perelli-Minetti: (805) 792-3162

OPEN DAILY 10 A.M.–5 P.M. WT, RS, P.

WP Red and White Table wines; Champagnes; Aperitifs; Dessert wines; Brandies   **R** Delano Elks Club   **HM** Stardust Motel.

MISC The Perelli-Minetti wine-making heritage dates back seven generations to eighteenth-century Milan. It was in 1902 that Antonio Perelli-Minetti brought his knowledge of wine making to America. Today his sons still follow family traditions in the town of Delano. This late dean of California viticulture developed and patented many new varietals grown on the ranch, the most recent of which, Perelli 101, will soon be bottled. A. Perelli-Minetti & Sons operates 2,500 acres of vines. The annual crushing capacity is 100,000 tons, with a storage capacity of 20 million gallons. There is a Brandy storage capacity of 150,000 small oak barrels. The winery features display cases and a boutique.

CASA DE FRUTA / (Fourteen miles east of Gilroy on U.S. 152) Mailing Address: 6680 Pacheco Pass Highway, Hollister, California 95023
                                                             (408) 842-9316

OPEN DAILY 8 A.M.–6 P.M.; CLOSED CHRISTMAS. WT, RS.

WP Zinfandel; Zinfandel Rosé; Gewürztraminer; Apricot; Pomegranate; Blackberry; Raspberry   **R** Casa de Fruta has three informal family restaurants at the vineyards   **HM** Casa de Fruta Motel   **HS** Mission San Juan Bautista; Pinnacles National Momument.

MISC The Casa de Fruta heritage of wine making, canning, and fruit growing goes back to 1903. The vineyards feature wine tasting, cheese tasting, dried fruit tasting and sales, a delicatessen, a Bar-B-Que Restaurant, overnight accommodations at the motel and in the travel park, picnic facilities, swimming, miniature golf, steam train ride, carousel, trout fishing, and dancing and live entertainment on Sundays and holidays.

24 THE TRAVELER'S GUIDE TO THE VINEYARDS OF NORTH AMERICA

CHALONE VINEYARD / P.O. Box 855, Stonewall Canyon Road, Soledad, California 93960 Peter Watson-Graff: (415) 441-8975

OPEN BY APPOINTMENT ONLY. RS.

GR Chardonnay; Pinot Noir; Pinot Blanc; Chenin Blanc; French Colombard **HS** Soledad Mission; Pinnacles National Monument.

MISC Founded in 1919, this small winery is now headed by Richard Graff and Phil Woodward.

B. CRIBARI & SONS WINERY / 3223 East Church Avenue, Fresno, California 93714 (209) 485-3080

OPEN DAILY 10 A.M.–5 P.M.; CLOSED NEW YEAR'S DAY, EASTER, THANKSGIVING, CHRISTMAS. WT, RS.

GR Red and White varietals **HS** Kearney Mansion

MISC The winery had its beginning in the early 1900s when Beniamino Cribari, a native of Italy, settled in Santa Clara county on forty acres of land. At the end of Prohibition a second Cribari winery was established on the third floor of a loft building in Manhattan, which became famous as "The Winery in the Sky." Today the tiny Cribari vineyards have grown to some five thousand acres under cultivation, and what started with a few bottles of homemade wine has grown to be a national business.

CONGRESS SPRINGS VINEYARDS / 23600 Congress Springs Road, Saratoga, California 95070 (408) 867-1409

OPEN SATURDAY AND SUNDAY NOON–5 P.M. WT, RS, P.

GR Zinfandel; Chardonnay; Gewürztraminer **R** La Mere Michelle; Plumed Horse; Le Mouton Noir; La Hacienda Inn **HM** Saratoga Motel; La Hacienda Inn **HS** Villa Montalvo Artists' Colony; Paul Masson Mountain Winery and Vineyard; Big Basin Redwood State Park.

MISC These vineyards, originally established in 1905, are the work of French immigrant Pierre C. Pourroy. Vic Erickson and Daniel Gehrs founded Congress Springs Vineyards in 1976. The winery specializes in Table wines from the eastern slopes of the Santa Cruz Mountains.

CYGNET CELLARS / 11736 Cienega Road, Hollister, California 95023 Jim Johnson: (408) 733-4276 (office); (408) 637-7559 (winery)

OPEN BY APPOINTMENT ONLY. WT, RS, P.

CALIFORNIA: THE CENTRAL COAST AND THE SAN JOAQUIN VALLEY

WP Cabernet Sauvignon; Zinfandel; French Colombard   **R** Tres Pinos Inn; Paines; Cutting Horse   **HS** Mission in San Juan Bautista.

MISC These cellars, founded in 1977, produce six thousand gallons of wine annually.

FELTON-EMPIRE VINEYARDS / 379 Felton-Empire Road, Felton, California 95018   William Gibbs: (408) 335-3939

OPEN BY APPOINTMENT ONLY. WT, RS, P.

GR White Riesling; Cabernet Sauvignon   **R** La Chaumiere; Santa Cruz Hotel Bar and Grill; Shadow Brook Restaurant   **HM** Pasatiempo Inn.

MISC The emphasis here is on traditional German wine-making methods for the White Rieslings, and Bordeaux methods for a limited production of Cabernet Sauvignon. Certain limited releases are available only at the winery.

FICKLIN VINEYARDS AND WINERY / 30246 Avenue 7½, Madera, California 93637   (209) 674-4598

NOT OPEN TO THE PUBLIC.

MISC David Ficklin and his son, Peter, inherited their love for wines from David's father, Walter C. Ficklin. David was directly involved in the construction of the original adobe building and has been the wine maker at the Ficklin Vineyards and Winery for thirty-one years. The winery specializes in Ficklin Port, made from a combination of Portuguese varieties. Ficklin is a small family operation, and not currently staffed to handle visitors, except by appointment, and then at only certain times of the yearly wine-producing cycle.

FIRESTONE VINEYARD / Zaca Station Road, Los Olivos, California 93441   (805) 688-3940

OPEN MONDAY–SATURDAY 10 A.M.–4 P.M.; CLOSED MAJOR HOLIDAYS. APPOINTMENT NECESSARY FOR GROUPS OF TEN OR MORE. WT, RS.

GR Cabernet Sauvignon; Merlot; Pinot Noir; Chardonnay; Johannisberg Riesling; Gewürztraminer   **R** Mattei's Tavern; Danish Inn; Ballard Store Restaurant.

MISC Founded in 1973, the Firestone Vineyard and Winery is still growing as its wine production increases.

FORTINO WINERY / 4525 Hecker Pass Highway, Gilroy, California 95020
Ernest Fortino: (408) 842-3305

OPEN DAILY 9 A.M.–6 P.M.; CLOSED CHRISTMAS. WT, RS, P.

GR Cabernet Sauvignon; Petite Sirah; Ruby Cabernet; Zinfandel; Sylvaner; Charbono; Carignane; Grenache; Grand Noir; Barbera **WP** Haut Sauterne; Chablis; Burgundy **R** Mt. Madonna Inn; Digger Dan's; Hecker Pass Inn **HM** Mount Madonna Inn **HS** Santa Clara City Museum.

MISC This winery originated in 1935. It was purchased in 1970 by Ernest Fortino, a third-generation wine maker born and raised in southern Italy. He converted the winery from bulk to varietal wine production. Over the years Fortino Winery has won gold medals for its fine California wines.

FRANZIA BROTHERS / 17000 East Highway 120, Ripon, California 95366
(209) 599-4111

TASTING ROOM OPEN DAILY 10 A.M.–5 P.M.; CLOSED NEW YEAR'S DAY, EASTER, THANKSGIVING, CHRISTMAS. WT, RS, P.

GR French Colombard; Palomino; Valdepenas; Zinfandel; Mission; Grenache; Ruby Cabernet; Chenin Blanc.

MISC In the 1976 Los Angeles County Fair Open Competition, Franzia Brothers wines received twenty-six awards. This prestigious family winery was founded in 1906 when Giuseppe Franzia, a native of Genoa, bought eighty acres of Ripon farmland. Since repeal of Prohibition Giuseppe's sons have been running the winery. James Walls, wine maker at Franzia Brothers, says, "Wine making has changed quite a bit since Giuseppe's days, but in some ways there are no shortcuts. It takes a blend of today's modern technology with the traditions of the old world to make a good wine."

THE FRICK WINERY / 303 Potrero Street, #39, Santa Cruz, California 95060
Judith M. Frick and William R. Frick: (408) 426-8623

OPEN BY APPOINTMENT ONLY. WT, RS.

GR Pinot Noir; Chardonnay; Gewürztraminer.

MISC The first and only winery in the city of Santa Cruz. Regular hours for tasting, tours, and retail sales. First release, '77 Chenin Blanc. '77 Pinot Noir also available.

CALIFORNIA: THE CENTRAL COAST AND THE SAN JOAQUIN VALLEY

GIUMARRA VINEYARDS / P.O. Box 1969, Bakersfield, California 93303   John Giumarra, Sr.: (805) 366-7253

OPEN WEDNESDAY–SUNDAY 9 A.M.–5 P.M.; CLOSED NEW YEAR'S DAY, EASTER, THANKSGIVING, CHRISTMAS. GROUPS BY APPOINTMENT. WT, RS.

GR All generics and varietals common to California **R** Banducci's **HS** Kern County Museum; Pioneer Village.

MISC This family-owned winery was founded in 1946. With a capacity of 10.5 million gallons, it is one of the largest in the United States. The bottling cellar operation can be seen from the tasting room. The family's vineyards are on the slopes of the Sierra Nevada Mountains adjoining the winery.

EMILIO GUGLIELMO WINERY / 1480 East Main Avenue, Morgan Hill, California 95037   Gene Guglielmo: (408) 779-2145

OPEN DAILY 9 A.M.–6 P.M. TOURS BY APPOINTMENT ONLY. WT, RS, P.

GR Riesling; Semillon; Grignolino; Gamay Beaujolais; Zinfandel; Petite Sirah; Ruby Cabernet **HS** Santa Clara Mission; Mission San Juan Bautista.

MISC Emilio Guglielmo, a Piedmontese Italian, used his life savings to purchase fifteen acres of fertile Santa Clara vineyard land in 1925. These vineyards have been in the family ever since.

HECKER PASS WINERY / 4605 Hecker Pass Highway, Gilroy, California 95020   Mario or Frances Fortino: (408) 842-8755

OPEN DAILY 9 A.M.–6 P.M.; CLOSED THANKSGIVING, CHRISTMAS. WT, RS, P.

GR Grenache; Petite Sirah; Zinfandel; Carignane; Ruby Cabernet.

MISC Mario Fortino learned the art of wine making in his family's winery in Cosenza, Italy. He and his wife have been at Hecker Pass since 1972, and their own establishment features a rustic tasting room.

HOFFMAN MOUNTAIN RANCH VINEYARD / Adelaide Road, Star Route, Paso Robles, California 93446   Sue Hoffman: (805) 238-4945
     Tasting Room: Corner of 24th Street and Black Oak, Paso Robles.

OPEN DAILY BY APPOINTMENT ONLY. TASTING ROOM OPEN DAILY 11 A.M.–5:30 P.M.; SUMMER HOURS 10:30 A.M.–6:30 P.M. WT AT TASTING ROOM ONLY.

Hoffman Mountain Ranch Vineyard (cont.)

GR Zinfandel; Chardonnay; Johannisberg Riesling; Cabernet Sauvignon; Pinot Noir; Sylvaner; Chenin Blanc; Grenache   **HM** Paso Robles Inn.

MISC In 1973 Dr. and Mrs. Hoffman established this winery where only estate-bottled premium wines are produced. That same year André Tchelistcheff, dean of California wine masters, joined the Hoffman Mountain Ranch as consultant. Through André's guidance in the vineyards and winery, sons David and Michael Hoffman produced wines which merited high honors at competitive tastings, including a gold medal at the 1975 International Wine Competition in London.

KIRIGIN CELLARS / 11550 Watsonville Road, Gilroy, California 95020
Nikola Kirigin Chargin: (408) 847-8827

OPEN DAILY 9 A.M.–6 P.M. GROUP TOURS BY APPOINTMENT ONLY. WT, RS, P.

GR Cabernet Sauvignon; Zinfandel; Pinot Noir; Malvasia Bianca; Sauvignon Vert   **R** Digger Dan's   **HS** Mount Madonna Park; Arthur Miller Ranch.

MISC Owner and wine maker Nikola Chargin is a native of Croatia with a degree in enology from the University of Zagreb. He came to the United States in 1959, and for several years he worked with other wineries, before establishing his own vineyards and cellars in 1976, when he purchased the former Bonesio Winery. The winery is built on the site of the Solis Rancho Homestead and incorporates a portion of the original building of 1827.

THOMAS KRUSE WINERY / 4390 Hecker Pass Road, Gilroy, California 95020
Tom Kruse: (408) 842-7016

OPEN WEEKENDS 10 A.M.–6 P.M.; WEEKDAYS BY APPOINTMENT ONLY. WT, RS; P LIMITED. FEE: 50¢ PER PERSON, REFUNDABLE WITH PURCHASE.

GR Cabernet Sauvignon; Chardonnay; Zinfandel   **R** Digger Dan's
HS Mount Madonna Park.

MISC Thomas Kruse, encouraged by several years of successful home wine making, purchased these vineyards in 1971 to begin production on a larger scale. The winery now produces its own bottle-fermented Champagnes in the classic *méthode champenoise*.

RONALD LAMB WINERY AND VINEYARDS / 17785 Casa Lane, Morgan Hill, California 95037
Ronald or Aldrene Lamb: (408) 779-4268

OPEN BY APPOINTMENT ONLY. RS.

CALIFORNIA: THE CENTRAL COAST AND THE SAN JOAQUIN VALLEY 29

WP Gamay Beaujolais; Amador Zinfandel; Cabernet Sauvignon; Napa Gamay; Chenin Blanc; Johannisberg Riesling; Chardonnay
HS Malaguerra Wine Museum.

MISC This winery was founded in 1975; its first crush was in 1976. The owners claim to run the smallest bonded winery in California.

LIVE OAKS WINERY / 3875 Hecker Pass, Gilroy, California 95020
Peter Scagliotti: (408) 842-2401

OPEN DAILY 8 A.M.–5 P.M.; CLOSED MAJOR HOLIDAYS. WT, RS, P.

GR Carignane; Mission; Golden Chasselas; Cabernet; Pinot Noir; Zinfandel **R** Harvest Time; Hecker Pass Inn; Digger Dan's **HS** Mount Madonna Park.

MISC It was 1912 when Eduardo Scagliotti purchased these vineyards. Today the Live Oaks Winery is still a family business. The winery specializes in "Premium Quality Burgundy," a blend of vintages aged for sixty-five months. Pure wine vinegar, both red and white, is also produced at Live Oaks.

LLORDS & ELWOOD WINERY / P.O. Box 3397, Fremont, California 94538 Information: 315 South Beverly Drive, Beverly Hills, California 90212
(213) 553-2368

OPEN BY APPOINTMENT ONLY.

WP Varietal Table wines; Champagne; Sherries; Port **R** Hugo's; Original Joe's; Plateau 7 **HM** Vagabond Motel; Hyatt House Hotel
HS International Rosicrucian Headquarters and Egyptian Museum; Lick Observatory; Winchester Mystery House.

MISC This is a small family winery established in 1955. The winery specializes in "late harvest" Johannisberg Riesling, Rosé of Cabernet Sauvignon, and the aging of Sherries by the Spanish method. Each of the wines has won numerous awards.

PAUL MASSON VINEYARDS / 13150 Saratoga Avenue, Saratoga, California 95070 Jack Welch: (408) 257-7800

OPEN DAILY 10 A.M.–4 P.M.; CLOSED NEW YEAR'S DAY, EASTER, THANKSGIVING, CHRISTMAS. WT, RS.

GR Chardonnay; Pinot Blanc; Johannisberg Riesling; Sylvaner; French Colombard; Semillon; Sauvignon Blanc; Cabernet Sauvignon; Gamay Beaujolais; Pinot Noir; Chenin Blanc; Gewürztraminer; Flora **R** La

Paul Masson Vineyards (cont.)

Hacienda; Los Gatos **HM** La Hacienda; Los Gatos **HS** Mission Santa Clara; Paul Masson Mountain Winery.

MISC Paul Masson, a native of Beaune, in Burgundy, inherited these vineyards from his father-in-law, who in turn had inherited them from *his* father-in-law, Etienne Thée, a vigneron of Bordeaux who had come to California in 1852. Paul Masson is the oldest continuous producer of wine in California, having operated during Prohibition under a special permit to produce wine for medicinal and sacramental use. This is the eighth largest winery in the United States with a total aging capacity of 28,220,371 gallons. Paul Masson is the largest exporter of premium wines from the United States: Masson wines are sold in over sixty countries. The vineyards offer a self-guided electronic tour system which gives wine information along a tour path. There is a retail store for wine and wine-related gifts.

MASTANTUONO VINEYARD / 101¾ Willow Creek Road, Paso Robles, California 93446   Pasquale Mastan: (805) 238-1078

OPEN ONLY BY APPOINTMENT.

GR Zinfandel.

MISC Founded in 1977, this new family-run vineyard is already producing six thousand gallons of Zinfandel annually.

**MIRASSOU VINEYARDS / 3000 Aborn Road, San Jose, California 95121
                                        Ruth Wiens: (408) 274-4000**

OPEN MONDAY–SATURDAY 10 A.M.–5 P.M.; SUNDAY NOON–4 P.M. CLOSED NEW YEAR'S DAY, EASTER, THANKSGIVING, CHRISTMAS. WT, RS; D FOR GROUPS BY APPOINTMENT.

WP Chardonnay; Dry Chablis; White Burgundy; Monterey Riesling; Johannisberg Riesling; Chenin Blanc; Gewürztraminer; Gamay Beaujolais; Burgundy; Zinfandel; Pinot Noir; Cabernet Sauvignon; Petite Sirah; Fleuri Blanc; Petite Rosé; Champagnes (*méthode champenoise*).

MONTEREY PENINSULA WINERY / 2999 Monterey–Salinas Highway, Monterey, California 93940   Barney Barnett: (408) 272-4949

OPEN DAILY 10 A.M. TO SUNSET; CLOSED CHRISTMAS. WT, RS, P.

WP Chardonnay; Zinfandel; Cabernet Sauvignon; Barbera; Pinot Noir; Zinfandel Essence; Plum; Apricot **R** Sardine Factory; Whaling Station Inn;

CALIFORNIA: THE CENTRAL COAST AND THE SAN JOAQUIN VALLEY

Consuello's; Stone House; Abalonetti Restaurant   **HM** Del Monte Hyatt House; Asilomar Conference Center; Highlands Inn; Ventyana Big Sur Inn; Casa Munras Hotel   **HS** Fisherman's Wharf; Cannery Row; Carmel Mission.

MISC Since this winery was founded in 1974 it has been honored with twenty-four medals from the Los Angeles County Fair, including seven gold medals. The winery has opened another wine tasting room, GIFTS FROM BACCHUS, in nearby Carmel.

MONTEREY VINEYARD / 800 South Alta Street, Gonzales, California 93926 (408) 675-2326

OPEN DAILY 10 A.M.–5 P.M.; CLOSED EASTER, CHRISTMAS. APPOINTMENT NECESSARY FOR GROUPS. WT, RS, P; D LUNCH ONLY.

R Sardine Factory; Whaling Station Inn; China Row; Quail Lodge; Highlands Inn; Gallatins   **HM** Del Monte Hyatt House; Monterey Hilton Inn; Monterey Holiday Inn   **HS** Soledad Mission; Pinnacles National Monument; Carmel Fort Ord; Steinbeck House.

MISC This vineyard now produces a total of eleven wines (all vintage-dated varietals) from six hundred acres in upper Monterey County. No generics are produced. Founded in 1974, the winery was purchased by the Coca Cola Company in 1977.

NICASIO VINEYARDS / 14300 Nicasio Way, Soquel, California 95073
Dan Wheeler: (408) 423-1073

OPEN SATURDAY 2 P.M. BY APPOINTMENT. WT, RS, P.

WP White Riesling; Chardonnay; Zinfandel Rosé; Zinfandel; Cabernet Sauvignon; Champagne   **R** Miramar; The Seascape   **HS** Santa Cruz Mission.

OBESTER WINERY / 12341 San Mateo Road, Half Moon Bay, California 94019 (415) 726-6465

OPEN SATURDAY AND SUNDAY 10 A.M.–5 P.M. WT, RS.

WP Zinfandel; Cabernet Sauvignon; Petite Sirah; Semillon; Sauvignon Blanc   **R** San Benito House.

MISC The Obester Winery is located in a picturesque valley where flowers and pumpkins are the main agricultural products. The winery is owned and operated by Paul and Sandy Obester and their teenage sons, and was inspired by Sandy's vintner grandfather, John Gemello.

PAPAGNI VINEYARDS / 31754 Avenue 9, Madera, California 93637
D. Lee Squyres: (209) 485-2760

OPEN MONDAY–FRIDAY 10 A.M.–3:30 P.M.; CLOSED ALL MAJOR HOLIDAYS. WT BY APPOINTMENT, RS.

GR Zinfandel; Barbera; Alicante Bouschet; Merlot; Grenache; Carignane; Charbono; Petite Sirah; Chardonnay; Chenin Blanc; Emerald Riesling; Sauvignon Blanc; Semillon; White Malaga   **R** Velvet Turtle; Refectory; Original Fondue and Crepe House; Bellante's; Pardini's; Royal Peking; Lucca's; Leilani   **HM** Picadilly Inn; Airport Marina; Holiday Inn; Smuggler's Inn; Ramada Inn   **HS** Kearney Mansion; Roeding Park; Fresno Arts Center; Madera Court House Museum.

MICHAEL T. PARSONS WINERY / 170 Hidden Valley Road, Soquel, California 95073   Mike Parsons: (408) 867-6070

OPEN BY APPOINTMENT ONLY. WT, RS.

GR Pinot Noir.

MISC Mike Parsons was a home wine maker for twelve years. A desire for sophisticated equipment and a new oak cooperage influenced his decision to bond this winery in 1976.

PEDRIZZETTI WINERY / 1645 San Pedro Avenue, Morgan Hill, California 95037   Ed Pedrizzetti: (408) 779-7380

OPEN DAILY BY APPOINTMENT; CLOSED NEW YEAR'S DAY, THANKSGIVING, CHRISTMAS. WT, RS.

GR Zinfandel; Barbera; Petite Sirah; Green Hungarian; Emerald Riesling.

MISC The Pedrizzetti Winery was established in 1945, and now three generations of the Pedrizzetti Family work the vineyards and the winery. There is an additional Pedrizzetti wine tasting room at 19020 Monterey, in Morgan Hill, open daily 10 A.M.–6 P.M. Call (408) 779-7774.

PESENTI VINEYARD / 2900 Vineyard Drive, Templeton, California 93465
A. Nerelli: (805) 434-1030

OPEN DAILY 8:30 A.M.–5:30 P.M.; CLOSED CHRISTMAS. WT, RS.

GR Zinfandel; Ruby Cabernet   **R** Iron Horse Restaurant   **HM** Motel 6   **HS** Hearst Castle.

MISC This third-generation vineyard was planted in 1923.

CALIFORNIA: THE CENTRAL COAST AND THE SAN JOAQUIN VALLEY 33

QUADY WINERY / 13181 Road 24, Madera, California 93637
Write to: Andrew Quady

OPEN SATURDAY AND SUNDAY 10 A.M.–4 P.M.; CLOSED ALL HOLIDAYS. RS.

WP Vintage Port   **R** Lucca's   **HM** Madera Valley Inn.

RANCHO SISQUOC WINERY / Route 1, Box 147, Santa Maria, California 93454
H. G. Pfeiffer: (805) 937-3616

OPEN BY APPOINTMENT ONLY. RS.

WP Cabernet Sauvignon; Rosé of Cabernet Sauvignon; Johannisberg Riesling   **HS** San Ramon Chapel at entrance to ranch.

MISC Rancho Sisquoc is a Spanish land grant of 3,800 acres. It is diversified into cattle, field crops, and vineyards, the latest addition being the winery.

MARTIN RAY WINERY / 22000 Mt. Eden Road, Saratoga, California 95070
Ken Brooks: (415) 321-6489

OPEN BY APPOINTMENT ONLY. RS.

GR Chardonnay; Cabernet Sauvignon; Pinot Noir; Riesling   **WP** As above plus Champagne from Chardonnay and Pinot Noir Blanc de Noir, *méthode champenoise*.

MISC This winery was founded in 1943 by the late Martin Ray, who had previously owned the Paul Masson Champagne Company. His son, Peter Martin Ray, is currently cellarmaster.

RICHERT & SONS WINERY / 1840 West Edmundson Avenue, Morgan Hill, California 95037
Scott Richert: (408) 779-5100

OPEN WEEKENDS 11 A.M.–4 P.M.; WEEKDAYS BY APPOINTMENT ONLY. WT, RS, P.

WP Sherry; Port; Fruit and Berry wines   **R** The Oak Tree Inn; Morgan House   **HM** Holiday Motel.

MISC Richert & Sons was founded in 1954 by Walter F. Richert. The winery is now managed by Scott Richert, an enology graduate of Fresno State College.

RIDGE VINEYARDS / 17100 Monte Bello Road, Cupertino, California 95014   David R. Bennion, president, Paul Draper, wine maker
(408) 867-3233

OPEN WEDNESDAY 11 A.M.–3 P.M. FOR SALES; OPEN SATURDAY 11 A.M.–3 P.M. FOR TASTING AND SALES. RS BY APPOINTMENT EXCEPT AS ABOVE.

GR Zinfandel; Cabernet Sauvignon; Petite Sirah.

MISC This winery is a renovation of the historic Montebello Winery, built deep into the rock by Osea Perrone in 1890. The vineyards and winery are a family project involving two generations of several families.

SANFORD & BENEDICT VINEYARD / Santa Rosa Road, Lompoc, California 93436   J. R. Sanford or M. R. Benedict: (805) 688-8314

OPEN BY APPOINTMENT ONLY. WT USUALLY; RS.

GR Pinot Noir; Chardonnay; Cabernet Sauvignon; Riesling; Merlot
R Mattei's Tavern; Danish Inn   **HM** Olifol.

MISC Planted in 1972, this 110-acre vineyard released its first wine in 1976. Sanford & Benedict specialize in Pinot Noir.

SAN MARTIN WINERY / P.O. Box 53, San Martin, California 95046
(408) 683-4000

TASTING ROOMS OPEN DAILY 10 A.M.–5 P.M.; CLOSED CHRISTMAS. WT, RS.

WP Pinot Chardonnay; Johannisberg Riesling; Emerald Riesling; Sauvignon Blanc; Chenin Blanc; Muscat di Canelli; Cabernet Sauvignon; Petite Sirah; Pinot Noir; Gamay Beaujolais; Zinfandel   **R** Digger Dan's.

MISC In 1973 wine maker Ed Friedrich came to what was then a small family winery to oversee a $1.5 million renovation of the production facilities. Under his direction San Martin Winery has been awarded many international medals as well as numerous awards in this country. Though the winery is not open to the public, there are four tasting rooms from which to sample San Martin's wares. They are: San Martin tasting room, Monterey Road, San Martin; 1110 San Pedro Avenue, Morgan Hill; Highway 25 and U.S. 101, Gilroy; and 475 Alisal Road, Solvang.

SANTA BARBARA WINERY / 202 Anacapa Street, Santa Barbara, California 93101   Pierre Lafond: (805) 962-3812

OPEN DAILY 10 A.M.–5 P.M. WT, RS.

CALIFORNIA: THE CENTRAL COAST AND THE SAN JOAQUIN VALLEY

GR Cabernet Sauvignon; Chenin Blanc; Zinfandel; Johannisberg Riesling; Pinot Chardonnay   **R** The Lobster House; The Chart House   **HS** Santa Barbara Mission; Santa Barbara Court House.

MISC Located one block from the beach, this winery produces both Fruit and Table wines.

SANTA YNEZ VALLEY WINERY / 365 North Refugio Road, Santa Ynez, California 93460   William H. Davidge: (805) 688-8381

OPEN BY APPOINTMENT ONLY. WT, RS.

GR Cabernet Sauvignon; Chardonnay; Gewürztraminer; Semillon; Sauvignon Blanc; Johannisberg Riesling; Merlot   **HS** Santa Ynez Mission; Historical Society Museum.

MISC The winery at Santa Ynez was established in 1976 on the site of the first college founded in California.

SHERRILL CELLARS / 1185 Skyline Boulevard, Palo Alto, California (no mail to this address). Mail to: P.O. Box 4155, Woodside, California 94062
         Jan W. or Nathaniel D. Sherrill: (415) 941-6023

OPEN BY APPOINTMENT ONLY. WT, RS.

WP Chardonnay; Zinfandel; Cabernet Sauvignon; Petite Sirah; Gamay.

MISC Originally located underneath the U.S. Post Office in Woodside, this husband-and-wife winery moved to its current location in 1979. There are invitational tastings once or twice yearly for those on the mailing list. Write to be included.

ROUDON SMITH VINEYARDS / 513 Mountain View Road, Santa Cruz, California 95065   Robert Roudon or James Smith: (408) 427-3492

OPEN BY APPOINTMENT ONLY. WT, RS.

GR Chardonnay   **R** La Trattoria; The Salmon Poacher; Pearl Alley.

SMOTHERS / 2317 Vine Hill Road, Santa Cruz, California 95065
         Linda Smothers: (415) 438-1260

OPEN BY APPOINTMENT ONLY.

GR White Riesling; Chardonnay.

Smothers (cont.)

MISC These vineyards date prior to 1900, and the winery was founded in February, 1977. Owner and television personality Dick Smothers is assisted by wine-making consultant Leo McCloskey in the limited production of estate-bottled and premium coastal California varietal wines.

SOMMELIER / 2560 Wyandotte Avenue, Section C, Mountview, California 94043   Mollie Keezer: (415) 969-2442 or 948-4379

OPEN SATURDAY AND SUNDAY 8 A.M.–5 P.M. BY APPOINTMENT ONLY. WT, RS.

WP Zinfandel; Cabernet Sauvignon; Grenache Rosé; Ruby Cabernet, Petite Sirah; Pinot Noir   **HM** Motel 6.

MISC The first wines were ready for market in 1978.

SUNRISE WINERY / 16001 Empire Grade Road, Santa Cruz, California 95060   Keith Hohlfeldt: (408) 423-8226
          Rolayne Stortz or Ronald Stortz: (408) 286-1418

OPEN BY APPOINTMENT ONLY. WT, RS, P.

GR Cabernet Sauvignon; Chenin Blanc; Pinot Chardonnay; Zinfandel
R Scopazzi's; Buffalo Gals; Santa Cruz Bar and Grill.

MISC The winery was founded in 1898 by the Locatelli Family; in 1976, Sunrise Winery took over the operation. The winery is built underneath the Locatelli Ranch House, and water is supplied by natural underground springs the year round.

SYCAMORE CREEK VINEYARDS / 12775 Uvas Road, Morgan Hill, California 95037   Terry Parks: (408) 779-4738

OPEN SATURDAY AND SUNDAY NOON–5 P.M. APPOINTMENT NECESSARY FOR TOURS. WT, RS, P.

GR Chardonnay; Zinfandel; Carignane   **R** Digger Dan's; Flying Lady; Mount Madonna Inn.

MISC Originally a pre-Prohibition winery, Sycamore Creek Vineyards was rebonded in 1976 by Terry and Marykaye Parks. New rootstock was planted and will be grafted into the Chardonnay grape.

CALIFORNIA: THE CENTRAL COAST AND THE SAN JOAQUIN VALLEY

LAS TABLAS WINERY / Winery Road, Templeton, California 93465
John and Della Mertens: (805) 434-1389
OPEN DAILY 9 A.M.–5 P.M. WT, RS, P.

GR Zinfandel   **WP** Zinfandel; Grenache Rosé; White Table wines; Sweet Muscat   **R** Paso Robles Inn; The Overland Stage; The Iron Horse   **HM** Paso Robles Inn; Black Oak Motel   **HS** San Miguel Mission; San Luis Obispo Mission; Hearst Castle.

MISC This winery was in the business of making and selling wine to early California settlers in 1856. Founded by Adolph Siot, the winery continued to prosper under the ownership of the Rotta Family. The Mertens Family purchased it in 1976.

VEGA VINEYARDS / 9496 Santa Rosa Road, Buellton, California 93436
William M. Mosby: (805) 736-7544
OPEN WEDNESDAY–MONDAY BY APPOINTMENT ONLY. WT, RS, P.

GR Johannisberg Riesling; Gewürztraminer; Sauvignon Blanc; Pinot Noir   **HS** La Purisma Mission; Santa Ynez Mission.

MISC The tasting room at this new, family-owned-and-operated winery is housed in an adobe house built in 1853 by one of the first families of the Spanish Land Grant era.

VENTANA VINEYARDS AND WINERY / P.O. Box G, Soledad, California 93960
J. Douglas Meador: (408) 678-2306
OPEN BY APPOINTMENT ONLY. WT, RS.

GR White Riesling; Gamay; Chenin Blanc; Zinfandel; Pinot Noir; Chardonnay; Petite Sirah; Cabernet Sauvignon   **R** The Sardine Factory; The Whaling Station; Shadow Brook   **HS** Soledad Mission.

NICHOLAS G. VERRY, INC. / 400 First Street, Parlier, California 93648
John N. Verry: (209) 646-2785
OPEN DAILY BY APPOINTMENT ONLY. WT.

WP Retsina.

MISC Nicholas Verry has been specializing in Grecian wines since 1933.

VILLA BIANCHI WINERY / 5806 Modoc, Kerman, California 93630
Max Weisner: (209) 846-7356

OPEN MONDAY–FRIDAY 8 A.M.–4 P.M. WT, RS.

GR Grenache; Cabernet Sauvignon; French Colombard; Thompson
R Picadilly   **HM** Picadilly Hotel   **HS** Winery Museum; Queen Mary Village.

YORK MOUNTAIN WINERY / Route 1, Box 191, Templeton, California 93465
Max Goldman: (805) 238-3925

OPEN DAILY 10 A.M.–5 P.M.; CLOSED HOLIDAYS. TOURS BY APPOINTMENT ONLY. WT, RS.

GR Chardonnay; Pinot Noir; Zinfandel; Cabernet Sauvignon   **R** Paso Robles Inn; The Iron Horse; Hamlin Restaurant   **HM** Paso Robles Inn
HS Hearst Castle; Morro Rock.

MISC It was to this winery, founded in 1882, that Ignacz Paderewski delivered grapes for his own wine. Purchased in 1970 by Max Goldman, a technologist in the industry since 1934, York Mountain plans to produce Champagne in the next few years in addition to the current seven varietals.

California: The East Bay Region and the Livermore Valley

Across the bay from San Francisco is Contra Costa, the "opposite coast." This is the route to the Sacramento and Lodi districts, passing through the wine country of the Livermore Valley on the way. If you are driving up from the south, on Interstate 5, you should go all the way into San Francisco (via Interstate 580 to Interstate 80 in Berkeley), to visit the San Francisco Wine Museum. This museum of wines, wine making, and wine drinking is a superb collection of wine-related art and historical pieces, including a fourth-century Roman wine cup, a Picasso, and much more. A visit makes a grand experience for wine lovers. The museum is open 11 A.M.–5 P.M. Tuesday through Saturday, and noon–5 P.M. on Sunday. The San Francisco Wine Museum is located at 633 Beach Street, opposite the Cannery at Fisherman's Wharf. From San Francisco, your next stop should be Wine for the People, at 907 University Avenue in Berkeley. This is a sampling room with an excellent reputation. While there, you might ask to taste the award-winning Gewürztraminer.

From Berkeley, Interstate 580 leads to the Livermore Valley, which is a fourteen-mile-long basin between the Hamilton and the Diablo mountain ranges. There are stores and restaurants in Livermore itself, and picnic tables at most of the valley's vineyards. Excellent wines, both Red and White, are made here, although it is the Whites that have gained the widest fame. This is plainer country than the lush valleys to the north, but the weather is essentially the same blend of cool nights and hot days. The wines here are well worth the traveling.

Interstate 580 out of Livermore to Interstate 5 going north will take you to Lodi, a temperate region with cool, moist winds, and one of the most important wine-growing and -processing centers in the state. Wine is the leading industry around Lodi, and every September Lodi hosts the biggest wine festival in America, the Lodi Grape Festival and National Wine Show. Lodi opens into the Sacramento Valley, where wine grapes have been grown since 1842. The valley is a distinguished wine district where vineyards appear in the middle of rich farmland. The area is dotted with places important in the history of the gold rush that took place in these parts. Sutter's Fort Historic Monument in Sacramento, for instance, is worth visiting in that regard.

If you then want to go back to San Francisco, you should visit the University of California at Davis, just outside the town of Davis on Interstate 80. The vineyards and experimental wine-making work at this famous agricultural school are open to the public. You can write for information to the Memorial Union Office, Freeborn Hall, University of California, Davis, California 95616.

ARGONAUT / Route 1, Box 612, Ione, California 95640
W. M. Bilbo: (714) 552-7149
OPEN BY APPOINTMENT ONLY. RS.

GR Barbera.

MISC Owned jointly by several aerospace engineers, this winery evolved out of W. M. Bilbo's interest in the art of home wine making. The first crush at Argonaut was in 1976.

BOEGER WINERY / 1709 Carson Road, Placerville, California 95667
Greg Boeger: (916) 622-8094
OPEN WEDNESDAY–SUNDAY 10 A.M.–5 P.M.; CLOSED MAJOR HOLIDAYS. WT, RS, P.

GR Cabernet Sauvignon; Merlot; Zinfandel; Chardonnay; Sauvignon Blanc; Flora; Semillon; Petite Sirah   **R** The Vineyard House   **HM** The Vineyard House   **HS** The Gold Bug Mine; Coloma Gold Discovery Site; Marshall Monument.

CALIFORNIA: THE EAST BAY REGION AND THE LIVERMORE VALLEY 41

MISC Placerville, known as Hang Town in the Gold Rush period, is experiencing a rebirth of vineyards. The tasting room at the Boeger Winery is located in a hundred-year-old winery listed in the National Register of Historic Places.

BRONCO WINE COMPANY AND JFJ WINERY / 6342 Bystrum Road, Ceres, California 95307   Fred T. Franzia: (209) 538-3131

OPEN BY APPOINTMENT ONLY. RS.

WP Sparkling and Table wines.   **R** Sundial Restaurant; Track 29; Cote d'Oro; Fisherman's Gallery   **HM** Sundial Lodge   **HS** McHenry Mansion and Museum.

MISC The Franzia Family has been producing California wines for three generations. It is the aim of the Bronco Wine Company to distribute the finest quality Northern California wines on a weekly basis throughout the state.

CADLOLO WINERY / 1124 California Street, Escalon, California 95320
(209) 838-2457

OPEN MONDAY–SUNDAY 9 A.M.–5 P.M.; CLOSED MAJOR HOLIDAYS. WT, RS.

WP Table and Dessert wines.

MISC Louis Sciaroni came to what is now Escalon from his native Switzerland in the late 1800s. With the exception of the Prohibition period, the Cadlolo Winery has been in operation ever since. The winery is housed in the original building which features a hospitality tasting room to welcome visitors.

CALIFORNIA CELLAR MASTERS / P. O. Box 478, Lodi, California 95240
Marlow E. Stark: (209) 368-6681

TASTING ROOMS OPEN MARCH–DECEMBER DAILY 10 A.M.–5 P.M. WT.

WP Varietal and generic Table wines; Sparkling, Dessert, and Fruit wines   **HM** Royal Host; El Rancho.

MISC Cellar Masters is comprised of five tasting rooms and two small wineries. The Coloma cellars, founded in 1849, were the work of Martin Allhoff, who had come to America drawn by the lure of Gold Rush adventure. The Gold Mine Winery, in Columbia, is another early mining site and now a historical state park. California Cellar Masters tasting rooms can be enjoyed at both these locations, as well as at Coloma Cellar 1, on U.S. 99, Lodi; Coloma Cellar 2, Old Town, Sacramento; Coloma Cellar 3, Cayucos; Coloma Cellar 4, Oakdale; and Coloma Cellar 5, Escalon.

RICHARD CAREY WINERY / 1688 Timothy Drive, San Leandro, California 94577   Richard Carey: (415) 352-5425

OPEN FRIDAY BY APPOINTMENT; SATURDAY 11 A.M.–4 P.M. WT, RS.

WP Zinfandel; Malvasia Bianca; Sauvignon Blanc; Chenin Blanc.

MISC Wine maker Richard Carey was an instructor at San Jose State University and had been an amateur wine maker for years. It was his hope to leave academia and fulfill his dream of making premium California wines. To this end he bought out the Chateau Vintners and established his own winery. His 1977 crush was the first at the Carey Winery, and although Carey explains that his is a small winery with a capacity of only ten thousand gallons, most of the work there is done by family and close friends.

CONCANNON VINEYARD / 4590 Tesla Road, Livermore, California 94550   Donna Wilcox: (415) 447-3760

OPEN MONDAY–SATURDAY 9A.M.–4 P.M.; SUNDAY NOON–4:30 P.M.; CLOSED NEW YEAR'S DAY, EASTER, THANKSGIVING, CHRISTMAS. TOURS WEEKDAYS ON THE HOUR, SATURDAY AND SUNDAY ON THE HOUR, NOON–3P.M. WT, RS, P.

GR Sauvignon Blanc; Muscat Blanc; Chenin Blanc; Johannisberg Riesling; Semillon; Franch Colombard; Petite Sirah; Cabernet Sauvignon   **R** La Rachelle's; Marchand's; Red Barron   **HM** Holiday Inn

MISC Born in Galway County in 1847, James Concannon emigrated to the United States as a youth. When he learned that the expanding archdiocese of San Francisco was in need of sacramental wine, he purchased land in Livermore Valley and planted these vineyards, which are now tended by his grandson James.

DELICATO VINEYARDS / 12001 South Highway 99, Manteca, California 95336   Dorothy Indelicato: (209) 239-1215 or 982-0679

OPEN DAILY 9 A.M.–5 P.M.; CLOSED NEW YEAR'S DAY, EASTER, THANKSGIVING, CHRISTMAS. TOURS BY APPOINTMENT. WT, RS, WP.

GR Carignane; Zinfandel; Grenache; Golden Chasselas   **R** Godfather's; Cancun; Packing Shed; Brass Door Restaurant; Bratskellar Restaurant; Croce's; Pietro's; Villa Basque; Alustiza's   **HM** Stockton Inn; Holiday Inn.

MISC Twenty members of the Indelicato Family now operate this winery, which features an antique wine press and a new round tasting room.

CALIFORNIA: THE EAST BAY REGION AND THE LIVERMORE VALLEY 43

DIABLO VISTA / 674 East H Street, Benicia, California 94510
Karen Borowski: (415) 837-1801

OPEN SATURDAY AND SUNDAY. WT, RS.

WP Cabernet Sauvignon; Zinfandel; Pinot Noir; Pinot Chardonnay
R Sam's Harbor; Commondant's **HM** North Bay TraveLodge **HS** Site of the first capital of California.

MISC Prior to the founding of this winery, the Borowskis were home wine makers. The first crush at Diablo Vista was in the fall of 1977.

EAST-SIDE WINERY / 6100 East Highway 12, Lodi, California 95240
Ernest C. Hass: (209) 369-4768

OPEN DAILY 9 A.M.–5 P.M.; CLOSED NEW YEAR'S DAY, THANKSGIVING, CHRISTMAS. WT, RS, P.

WP Dessert and Table wines; Brandy **R** Croce's; John's Back Bay; Golden Acorn.

MISC The tasting room at the East-Side Winery is set within a fifty-thousand-gallon redwood tank.

ELDORADO VINEYARDS / 3551 Carson Road, Camino, California 95709
Mick or Jody McGuire: (916) 644-3773

OPEN DAILY 9 A.M.–6 P.M. LABOR DAY–CHRISTMAS; OPEN WEEKENDS AND BY APPOINTMENT REST OF YEAR. WT, RS, P.

GR Zinfandel; Chenin Blanc; White Riesling; Cabernet Sauvignon; Chardonnay; Merlot; Sauvignon Blanc; Apple and other Fruit wines
R Gringo's; Arnavais **HM** Mother-Lode Motel; Gold Trail Motel
HS Gold Discovery Park; Gold Bug Mine; Eldorado County Museum.

MISC In 1973 the McGuires purchased the Dodd Ranch and Homestead with accompanying apple orchard in order to develop a small family winery. The ranch had been neglected, but the McGuires are beginning to bring the thirty acres to full production with apples and wine grapes. Apples and wines may be purchased on the premises. Visitors are invited to observe the apple harvest in September.

GALLO VINEYARDS / P. O. Box 1130, Modesto, California 95353
Charles M. Crawford: (209) 521-3111

OPEN BY APPOINTMENT ONLY.

Gallo Vineyards (cont.)

GR Sauvignon Blanc; Chenin Blanc; French Colombard; Ruby Cabernet; Barbera; Zinfandel.

MISC Founded in 1933 by Ernest and Julio Gallo, the winery now has a storage capacity of over 226 million gallons with wineries in Modesto, Livingston, Fresno, and Healdsburg. A visitors' tasting room is planned for the future but is as yet unscheduled.

THE GEMELLO WINERY / 2003 El Camino Real, Mountain View, California 94040
Alec White: (415) 948-7723

OPEN SATURDAY NOON–4 P.M. OR BY APPOINTMENT. WT, RS.

WP Cabernet Sauvignon; Zinfandel; Petite Sirah; Pinot Noir; Gamay Beaujolais; Chenin Blanc **R** La Terrace; Liaison; Onion Barn **HM** Hyatt Cabana; Holiday Inn; Tropicana Lodge.

GIBSON WINE COMPANY / 9750 Kent Street, Elk Grove, California 95624
Tasting room: Highway 99 & Grantline Road   (916) 685-9594

WINERY NOT OPEN TO PUBLIC. TASTING ROOM OPEN DAILY. WT, RS, P.

GR Thompson; Colombard; Grenache; Carignane; Ruby Red **R** Destiny Restaurant **HM** Holiday Inn South **HS** Elk Grove Park.

MISC What started as a privately owned winery in 1934 is now a growers' cooperative with 140 members. The Gibson Wine Company ships approximately 1.2 million cases of wine per year to forty states.

HARBOR WINERY / 7576 Pocket Road, Sacramento, California 95831
Charles H. Myers: (916) 392-7954

NOT OPEN TO THE PUBLIC.

WP Chardonnay; Zinfandel; Cabernet Sauvignon; Mission Del Sol.

MISC Founded in 1972, the Harbor Winery sells wine to exclusive restaurants, including the Ritz Hotel in London.

MONTCLAIR WINERY / 910 81st Avenue, Oakland, California
(415) 658-1014

NOT OPEN TO THE PUBLIC.

CALIFORNIA: THE EAST BAY REGION AND THE LIVERMORE VALLEY 45

WP Zinfandel; Cabernet Sauvignon; French Colombard.

MISC Originally known as the Big Foot Winery, Montclair is still one of the smallest wineries in California. Owner R. K. Dove comments: "Our commitment is not growth oriented, but rather to produce wines to our taste, which we hope you will enjoy. If you wish to receive our newsletter, drop us a note at 180 Maxwelton Road, Piedmont, California 94618."

MONTEVINA WINERY / Route 2, Box 30-A, Plymouth, California 95669
W. H. Field or T. C. Gott: (209) 245-3412

OPEN MONDAY–FRIDAY 9 A.M.–5 P.M.

GR Zinfandel; Nebbiolo; Cabernet Sauvignon; Sauvignon Blanc; Mission; Ruby Cabernet; Merlot; Barbera; Chardonnay **R** Argonaut Inn.

J. W. MORRIS PORT WORKS / 1215 Park Avenue, Emeryville, California 94608
John Tarman: (415) 655-3009

OPEN BY APPOINTMENT ONLY. WT, RS.

WP Vintage Port; Table wines **R** Narsai's; The Bay Wolf **HS** Old Emeryville City Hall.

MISC This young winery, founded in 1975, recently won a gold medal at the Los Angeles County Fair for a 1975 vintage Port. This Port also won a gold medal at the International Wine and Cheese Show in Washington, D.C.

MOUNT EDEN VINEYARDS / 22020 Mt. Eden Road, Saratoga, California 95070
Bill Anderson: (408) 867-5783

OPEN BY APPOINTMENT ONLY. RS.

GR Chardonnay; Cabernet Sauvignon; Pinot Noir **R** Plumed Horse
HM Los Gatos Inn; La Hacienda.

MISC The Mount Eden Vineyards' first vintage year was 1972.

NOVITIATE WINES / 300 College Avenue, Los Gatos, California 95030
Brother Korte: (415) 354-6471

OPEN MONDAY–SATURDAY 9 A.M.–4 P.M.; CLOSED CHURCH AND LEGAL HOLIDAYS. TOURS MONDAY–FRIDAY AT 1:30 AND 2:30; SATURDAY AT 10 AND 11. WT, RS; P MAY–OCTOBER; D FOR SPECIAL GROUPS.

Novitiate Wines (cont.)

GR Pinot Blanc; Pinot Noir; Cabernet Sauvignon; Chenin Blanc; Semillon Malvasia; Riesling; Ruby Cabernet; Barbera; Grenache; Valdepenas; Zinfandel; Palomino.

MISC Two problems faced the earliest Jesuit Fathers and Brothers in 1898: how could they support the Novices training to become Jesuits, and where could they obtain valid Mass wine? Altar wine must be between 12 and 18% alcohol by volume, and be made from nothing but grapes if it is to be legitimate sacramental wine. The solution was carried out when Brother Louis Olivier, a vintner from southern France, sent home to Mt. Pelier for cuttings of the best varietals. The tradition started by Brother Olivier has been handed down through the generations to produce the Novitiate Wines available today.

OAK BARREL WINERY / 1201 University Avenue, Berkeley, California 95702   John Bank: (415) 849-0400
OPEN MONDAY–SATURDAY 10 A.M.–6:30 P.M. WT, RS.

WP Table wines.

MISC The Oak Barrel Winery is one of the largest suppliers of wine-making equipment in the country. They will carve barrels to order.

PAGE MILL WINERY / 13686 Page Mill Road, Los Altos Hills, California 94022   Ome or Dick Stark: (415) 948-0958
OPEN BY APPOINTMENT ONLY. WT, RS.

GR Chardonnay; White Riesling; Zinfandel   **HM** Rickey's Hyatt House; Holiday Inn.

MISC Owner and wine maker Dick Stark, who describes himself as "an escapee from the electronics industry," runs his winery with the help of his wife and four children. The first crush, in 1976, yielded one thousand gallons, and the 1977 crush yielded three times as much.

SEQUOIA CELLARS / 1101 Lincoln Avenue, Woodland, California 95695
         Patricia S. Riley: (916) 487-3210
NOT OPEN TO THE PUBLIC.

WP Zinfandel; Cabernet Sauvignon; Chenin Blanc; Gewürztraminer; Carnelian   **HS** Woodland Opera House.

CALIFORNIA: THE EAST BAY REGION AND THE LIVERMORE VALLEY 47

MISC Founded in 1977, this small but growing winery has an inventory.

SHENANDOAH VINEYARDS / Route 2, Box 23, Steiner Road, Plymouth, California 95669   Leon or Shirley Sobon: (209) 245-3698

OPEN BY APPOINTMENT ONLY. WT, RS.

GR Zinfandel; Sauvignon Blanc   **R** Mt. Aukum Lodge; Buscaglia's; Wheel Inn.

MISC These vineyards plan to remain a family-run operation. Their first crush (twenty-six tons) was in 1977. Fifty-six tons were crushed in 1978.

SIERRA VISTA / 45 Cabernet Way, Placerville, California 95667
                    John MacCready: (916) 622-7221

OPEN BY APPOINTMENT ONLY. WT, RS.

GR Cabernet Sauvignon   **WP** Cabernet Sauvignon; Zinfandel; White Zinfandel; Rosé of Cabernet Sauvignon; Mother Lode Gold   **HS** Gold Discovery Site.

MISC Vineyard established in 1974; winery bonded in 1977.

STONERIDGE / Route 1, Box 36-B, Ridge Road East, Sutter Creek, California 95685   Gary or Loretta Porteous: (209) 223-1761

OPEN SATURDAY 2 P.M.–4 P.M. AND BY APPOINTMENT. WT, RS.

GR Zinfandel; Ruby Cabernet   **R** Buscaglia's; Suraci's; Teresa's; Wheel House; Sutter Creek Palace; Belotti's; Jug and Rose   **HM** St. George Hotel; Linda Vista; Holiday Inn; El Campo Casa   **HS** Amador County Museum; Sutter Creek Grammar School; Knights Foundry; Chaw-se Indian Grinding Rocks State Historic Park and Mi-wok Cultural Center; Kennedy Mine Tailing Wheels Park

MISC The vineyards were established in 1972 and bonded in 1975.

STONY RIDGE WINERY / 1188 Vineyard Avenue, Pleasanton, California 94566   Harry Rosingana: (415) 846-2133

OPEN 11 A.M.–5 P.M.: SEPTEMBER–MAY, WEEKENDS ONLY; JUNE–AUGUST, TUESDAY–SUNDAY. WT, RS, P.

GR Zinfandel; Riesling; Malvasia Bianca; Chardonnay   **R** Hacienda del Sol; Emperor's Garden   **HM** Holiday Inn; Howard Johnson's.

Stony Ridge Winery (cont.)

MISC In 1975 Harry and Len Rosingana restored the Hillside Winery and Vineyards which had been established in 1887 by Lou Crellin. The Rosinganas comment: "The intense varietal character of our wines is the result of strict viticultural practices of thinning and suckering the crop to produce less than one ton per acre."

TURGEON AND LOHR WINERY / 1000 Lenzen Avenue, San Jose, California 95126   Bernie Turgeon or Jerry Lohr: (408) 288-5057

OPEN DAILY 10 A.M.–5 P.M. WT, RS, P.

GR Cabernet Sauvignon; Petite Sirah; Zinfandel; Pinot Noir; Gamay Beaujolais; Merlot; Napa Gamay; Johannisberg Riesling; Pinot Blanc; Sauvignon Blanc; Chardonnay   **R** Le Baron; Angello's; The Monks Retreat; 94th Arrow Squadron; Mama Bongiorno's; The Plumed Horse; Le Mouton Noir; Mountain Charlie's   **HM** Le Baron Hotel; Holiday Inn; Hyatt House   **HS** Santa Clara Mission; The Rosicrucian Museum; Winchester Mystery House; Villa Montalvo.

MISC Deciding to return to their farming roots, Jerry Lohr and Bernie Turgeon planted these vineyards in 1972. Soon joined by master enologist Peter Stern, they opened their winery in San Jose.

VEEDERCREST WINERY / 1401 Stanford Avenue, Emeryville, California 94608   (415) 652-3103

VISITS BY APPOINTMENT ONLY. WT; RS BY CASE ONLY.

WP Merlot; Cabernet Sauvignon; Chardonnay; White Riesling; Gewürztraminer; Blanc de Pinot Noir   **R** Trader Vic's.

MISC Although comparatively small and quite new, Veedercrest has taken a respectable place among a new generation of internationally oriented small wineries. A string of important medals from the Los Angeles County Fair and acclaim from San Francisco's Vintners Club attest to the quality of these wines. Recognition in important blind tastings in Paris and Beaune have established the high quality of Veedercrest wines in the Claret and White Burgundy families.

CONRAD VIANO WINERY / 150 Morello Avenue, Martinez, California 94553   (415) 228-6465

OPEN DAILY 9 A.M.–NOON, 1 P.M.–5 P.M.; CLOSED MAJOR HOLIDAYS. WT, RS.

CALIFORNIA: THE EAST BAY REGION AND THE LIVERMORE VALLEY 49

GR Wine grapes   **R** The Warehouse Cafe   **HS** John Muir Home; Mount Diablo.

VILLA ARMANDO / 553 St. John Street, Pleasanton, California 94566
Dominic Scotto: (415) 846-5488

OPEN DAILY NOON–5:30 P.M.; CLOSED NEW YEAR'S DAY, EASTER, THANKSGIVING, CHRISTMAS. WT, RS, D.

GR Red Rustico 16%; White Rustico; Rustico Rosé; Pinot Blanc; Ruby Cabernet; Cabernet Sauvignon; Petite Sirah   **R** La Villa Armando   **HM** Howard Johnson's; Holiday Inn   **HS** Mount Diablo; Lawrence Radiation Laboratories.

MISC The Scotto Family has been making wines for over one hundred years. Leaving their native Bay of Naples in 1903, they settled in Pleasanton and have been making American wines on this land ever since. Villa Armando is presently expanding its facilities in order to expand its sales throughout the United States.

WENTE BROTHERS / 5565 Tesla Road, Livermore, California 94550
Lawrence L. DiPietro: (415) 447-3603

OPEN MONDAY–SATURDAY 9 A.M.–5 P.M.; SUNDAY 11 A.M.–4 P.M.; CLOSED NEW YEAR'S DAY, EASTER, THANKSGIVING, CHRISTMAS. APPOINTMENT REQUIRED FOR BUS TOURS. WT, RS, P.

GR Pinot Chardonnay; Grey Riesling; Chenin Blanc; Sauvignon Blanc; Semillon; French Colombard; Pinot Blanc; Pinot Noir; Zinfandel; Gamay Beaujolais; White Riesling; Sylvaner; Petite Sirah; Ugni Blanc   **HM** Holiday Inn.

MISC Carl Wente started his family winery in the Livermore Valley in 1883. Today the fourth generation actively carries on his tradition.

WINE AND THE PEOPLE / 907 University Avenue, Berkeley, California 94550
(415) 549-1266

OPEN DAILY 10 A.M.–6 P.M.; CLOSED MAJOR HOLIDAYS. SUNDAY TOURS BY APPOINTMENT ONLY. WT, RS.

WP Merlot; Cabernet Sauvignon; Pinot Noir; Chardonnay; Zinfandel; Zinfandel Port   **R** Omnivore; Chez Panisse; Narsai's; Bay Wolfe; Spenger's   **HM** Marriott Inn; Berkeley House; Claremont Hotel
HS Berkeley Museum; Oakland Museum; San Francisco Wine Museum.

Wine and the People (cont.)

MISC Wine and the People specializes in classic Dry wines in the French style. Wine maker Peter Brehm was awarded two gold medals at the 1974 California State Fair, and his Gewürztraminer was declared Best Wine of the Exposition.

WINEMASTERS WINERY / 1 Winemasters Way, Lodi, California 95240
(209) 368-5151

OPEN DAILY 10 A.M.–5 P.M.; CLOSED NEW YEAR'S DAY, EASTER, THANKSGIVING, CHRISTMAS. WT, RS, P; D FOR PRIVATE GROUPS OF TWELVE BY RESERVATION ONLY.

GR Red and White wine varietals   **HS** Mariposa Big Tree Grove; Mother Lode Gold Country.

MISC The Winemasters brand offers American consumers a highly diversified line of California and European wines under one distinctive label. Through this unique marketing approach, wines are carefully selected by a group of wine masters in California and in the wine-producing regions of Europe. The line features California varietal and generic Table wines, Sparkling and Dessert wines, as well as selected European Table wines. The winery features a Brandy-aging warehouse and a display vineyard.

THE WINE MUSEUM OF SAN FRANCISCO / 633 Beach Street, San Francisco, California 94109   Ernest G. Mittelberger: (415) 673-6990

OPEN TUESDAY–SATURDAY 11 A.M.–5P.M.; SUNDAY NOON–5 P.M.; CLOSED MONDAY AND MAJOR HOLIDAYS. DOCENT TOURS EVERY AFTERNOON 2 P.M.–4 P.M. ADMISSION FREE.

HM Holiday Inn; Sheraton Hotel; Ramada Inn; TraveLodge
HS Fisherman's Wharf; Maritime Museum; Ghirardelli Square.

MISC Unlike our other entries, the Wine Museum of San Francisco cannot offer wine tastings or scenic picnic areas. But the treasures it does offer, always free of charge, are too fascinating to pass up. Art works in The Wine Museum were assembled over a period of thirty-five years at the direction of Alfred Fromm, Chairman of the Board of Fromm & Sichel, Inc., exclusive worldwide distributors of The Christian Brothers wines and brandies. The Collection toured the country as a traveling exhibition at major museums until it found its permanent home opposite The Cannery on Fisherman's Wharf in 1974. The Museum offers: The Franz W. Sichel Glass Collection, a two-thousand-year history of the blown drinking glass; five hundred years of wine in the graphic arts; a rare wine book library of approximately one thousand volumes, dating from 1550 to the present; sculpture; and decorative arts.

CALIFORNIA: THE EAST BAY REGION AND THE LIVERMORE VALLEY 51

WOODEN VALLEY WINERY / Suisun Valley Road, Suisun, California 94584
(707) 864-0730

OPEN TUESDAY–SUNDAY 9 A.M.–5 P.M.; CLOSED MAJOR HOLIDAYS. WT, RS.

GR Riesling; Sauvignon Blanc; Golden Chasselas; Zinfandel; Pinot Noir; Cabernet Sauvignon; Gamay Beaujolais; Carignane; Early Burgundy.

MISC The Lanza Family founded Wooden Valley in 1932. At that time the annual capacity was three thousand gallons; at present, the winery's capacity is thirty-five times as great.

WOODSIDE VINEYARDS / 340 Kings Mountain Road, Woodside, California 94062
Robert L. Mullen: (415) 851-7475

OPEN WEEKENDS BY APPOINTMENT. WT, RS.

GR Cabernet Sauvignon; Pinot Noir; Chardonnay   **R** Stage Coach; Village Pub.

MISC The Cabernet Sauvignon grapes grown at Woodside were planted in 1900 by the Rixford Vineyards. All other grapes were planted by the Mullen Family, starting in 1961.

YANKEE HILL WINERY / P. O. Box 163, Columbia, California 95310
(209) 532-3015

OPEN DAILY SUNUP TO SUNDOWN. WT, RS, P.

WP Table, Fruit, and Sparkling wines   **R** Stage Coach Inn; City Hotel; Hungry Drifter   **HM** City Hotel; Columbia Inn   **HS** Gold Rush town of Columbia is now a state park.

MISC Ron Erickson established the Yankee Hill Winery in 1971.

California:
The Napa Valley

The Napa district is some fifty miles north of San Francisco. It is a beautiful valley stretching into wooded mountains that are volcanic in origin. The valley floor rises north toward the town of St. Helena in stages beginning with the first long meadows on San Francisco Bay. Cool, wet bay winds wash up the valley at night. Days are hot, and evening fogs creep in from the ocean. The Napa River drains the valley, which is a manifestly ideal place to grow wine grapes.

On the eastern side of the valley the Silverado Trail (a modern road) runs by the fields down the slopes to the bay. This is an elevated drive from which you can see the vineyards; and this lovely route is connected with the valley by several crossroads. Route 29, which intersects the length of the valley, is known as the "Wine Road," because in places it boasts a vineyard or winery at the rate of one per mile.

Picnicking is possible as far south in the valley as Yountville, and of course at vineyards. Many of the vineyards and wineries can provide you with cheeses and other good picnic foods. There are stores of all kinds in St. Helena, the hub of commercial life in the valley, and three-fourths of Napa's famous wineries are located within ten miles of the town. Hundred-year-old buildings line Main Street, whose street lamps were erected in 1916. Along with commercial establishments, St. Helena offers the only public wine library in the United States, the St. Helena Library. Farther north in the valley, Calistoga has hot springs, geysers, and a forest of petrified redwoods. Old Faithful geyser is a few miles north of Calistoga.

Napa Valley residents are very aware that they live in California's premier wine area. Napa wines are known in Europe, and French interests have been buying land around Napa. Though proud, the residents are not inhospitable. Highly conscious too of the popularity of their products, many of the historical-landmark wineries offer cellar tours and other special attractions—catered dinners for up to 150 people, and even an aerial tramway.

From San Francisco, the drive to the town of Napa takes about an hour. You may go on Interstate 80 to Vallejo and enter the valley on Route 29. An alternative suggestion is to drive on Route 37 past Ignacio onto Route 121 to Napa. From Route 101 onto Route 121 to Napa. From Route 101 in the Sonoma country, you can take Route 128 into Calistoga at the north end of Napa Valley. Experienced travelers caution that it is difficult to get reservations, and you are advised to phone ahead wherever you plan to visit. The St. Helena (wine) Library closes at 7 P.M. weekdays and 3 P.M. Saturdays (no reservations).

BEAULIEU VINEYARD / 1960 St. Helena Highway, Rutherford, California 94573 Peter Dahl: (707) 963-2411

OPEN DAILY 10 A.M.–4 P.M.; CLOSED NEW YEAR'S DAY, EASTER, THANKSGIVING, CHRISTMAS. WT, RS.

GR Cabernet Sauvignon; Pinot Noir; Chardonnay; Johannisberg Riesling.

MISC Founded in 1900 by Georges de Latour, a native of France's Perigord region, the Beaulieu Vineyard was family owned and operated until 1969, when it was sold to the Heublein Corporation. June 1979 marks the tenth anniversary of Heublein ownership. The winery features a fifteen-minute audio-visual presentation for visitors.

JOHN BECKETT CELLARS / 1055 Atlas Peak Road, Napa, California 94558 John Beckett: (707) 224-2022

CELLARS OPEN TO THE PUBLIC BY APPOINTMENT ONLY.

WP Cabernet Sauvignon; Fumé Blanc; Johannisberg Riesling.

CALIFORNIA: THE NAPA VALLEY

BERINGER VINEYARDS / 2000 Main Street, St. Helena, California 94574
Jean Manning: (707) 963-7115

OPEN DAILY 9:30 A.M.–4:30 P.M.; CLOSED NEW YEAR'S DAY, CHRISTMAS. APPOINTMENT NECESSARY FOR GROUPS. WT, RS.

GR Chardonnay; Pinot Noir; Grey Riesling; Chenin Blanc; Zinfandel; Petite Sirah; Napa Gamay; Johannisberg Riesling; French Colombard; Grignolino; Sauvignon Blanc; Semillon; Cabernet Sauvignon; Merlot   **HM** Wine Country Inn; Burgundy House; Magnolia Inn; Hotel Miramonte   **R** Mama Nina's; La Belle Helene; Silverado Saloon; Oakville Public House; The Abbey.

MISC In 1876 Jacob and Frederick Beringer established what is today the oldest continuously operating winery in the Napa Valley. In 1883 they built a seventeen-room mansion known as the Rhine House. The Rhine House now serves Beringer Vineyards as a hospitality and tasting center. The tour of the winery features a film, a walk through the underground cellars a thousand feet below the Rhine House, and ends with wine tasting in this historic building.

BURGESS CELLARS / P. O. Box 282, St. Helena, California 94574
Tom Burgess: (707) 963-4766

OPEN BY APPOINTMENT ONLY. RS, P.

GR Cabernet Sauvignon   **WP** Cabernet Sauvignon; Zinfandel; Pinot Noir; Chardonnay; Chenin Blanc; Johannisberg Riesling.

CAKEBREAD CELLARS / 8300 St. Helena Highway, Rutherford, California 94573   Jack Cakebread; (707) 963-9182 or (415) 835-WINE

OPEN BY APPOINTMENT ONLY. RS.

GR Sauvignon Blanc; Cabernet Sauvignon; Chardonnay; Zinfandel.

CARNEROS CREEK WINERY / 1285 Dealy Lane, Napa, California 94558
Francis V. Mahoney: (707) 226-3279

OPEN MONDAY–FRIDAY 9 A.M.–4 P.M. BY APPOINTMENT ONLY; CLOSED NEW YEAR'S DAY, THANKSGIVING, CHRISTMAS. RS.

GR Chardonnay; Pinot Noir   **R** The Carriage House.

MISC The Carneros Creek Winery was founded in 1971.

CASSAYRE-FORNI CELLARS / 1271 Manley Lane, Rutherford, California 94573
Sarah Forni: (707) 944-2165

OPEN BY APPOINTMENT ONLY. RS.

WP Cabernet Sauvignon; Zinfandel; Chenin Blanc.

MISC The owners of the Cassayre-Forni Cellars are third-generation Napa Valley residents whose interest in wine making has grown from a combination of knowledge gained while designing wineries with their engineering company and an inherited family involvement in the wine industry. The winery's first release was in the summer of 1978.

CAYMUS VINEYARDS / 8700 Conn Creek Road, Rutherford, California 94573
(707) 963-4202

OPEN MONDAY–SATURDAY 9 A.M.–5 P.M. BY APPOINTMENT ONLY. WT, RS.

GR Cabernet Sauvignon; Pinot Noir; Zinfandel; Burger; Johannisberg Riesling; Napa Gamay.

CHARLES KRUG WINERY / 2800 Main Street, St. Helena, California 94574
John F. Aquila: (707) 963-2761

OPEN DAILY 10 A.M.–4 P.M.; CLOSED MAJOR HOLIDAYS. APPOINTMENT NECESSARY FOR LARGE GROUPS. WT, RS.

GR Varietals.

MISC Registered as Historical Landmark 563, this winery was founded in 1861 by Charles Krug. Since 1943 the winery has been owned and operated by C. Mondavi and Sons.

CHÂTEAU CHEVALIER WINERY / 3101 Spring Mountain Road, St. Helena, California 94574
C. Gregory Bissonette: (707) 963-2342

OPEN BY APPOINTMENT ONLY. RS.

GR Cabernet Sauvignon; Merlot; Chardonnay; Pinot Noir; White Riesling   **HM** The Country Inn.

MISC The Chevalier vineyards and winery have been producing wines for over one hundred years. During this time the wines produced here have been the result of both careful attention and, in the hands of some owners, neglect. Fortunately the present owners, Gregory Bissonette, James Frew, and Peter Hauschildt, purchased the château in 1969, and have brought vigor and excitement to the restoration of this handsome estate.

CALIFORNIA: THE NAPA VALLEY

CHRISTIAN BROTHERS / 2555 Main Street, St. Helena, California 94574
(707) 963-2719

OPEN DAILY 10:30 A.M.–4:30 P.M.; CLOSED NEW YEAR'S DAY, GOOD FRIDAY, EASTER, THANKSGIVING, CHRISTMAS. WT, RS.

HS Old Bale Mill.

MISC The winery was established in 1882 by a teaching order of the Roman Catholic church. Winery sales help support thirteen schools in their western U.S. province. This is the largest stone winery in the world, with a two-million-gallon capacity.

CLOS DU VAL / 5330 Silverado Trail, Napa, California 94558
Bernard M. Portet: (707) 252-6711

OPEN WEEKDAYS 10 A.M.–4 P.M.; CLOSED MAJOR HOLIDAYS. RS.

GR Cabernet Sauvignon; Merlot; Zinfandel   **R** Domaine Chandon
HM Silverado Country Club.

MISC Clos de Val celebrated its first vintage in 1972. Wine maker Bernard Portet is the son of the *régisseur* of the famous Château Lafite Rothschild. While Bernard Portet maintains that the wines of California and France are distinct, and that comparison is not useful, he is trying to make his wines as close in character to the wines of Pauillac as possible.

CONN CREEK WINERY / 3222 Ehlers Lane, St. Helena, California 94574
William Collins: (707) 963-3945

OPEN BY APPOINTMENT ONLY. WT, RS, P.

GR Chardonnay; Zinfandel; Cabernet Sauvignon   **R** Domaine Chandon; La Belle Helene; Freemark Abbey   **HM** Wine Country Inn; El Bonita
HS Old Bale Mill.

MISC Although the Collins and Beaver families rebuilt the interior of the winery building, they are still using vines first planted in 1923.

CUVAISON / 4550 Silverado Trail, Calistoga, California 94515
Beverly Johnston: (707) 942-6266

OPEN THURSDAY–MONDAY 10 A.M.–4 P.M.; CLOSED NEW YEAR'S DAY, EASTER, THANKSGIVING, CHRISTMAS. RS, R.

WP Chardonnay; Cabernet Sauvignon; Zinfandel.

Cuvaison (cont.)

MISC In 1970 Cuvaison began as a small, do-it-yourself winery, based in an old farmhouse. But by early 1973 construction commenced on the present facilities. Under the guidance of Philip Togni, renowned Swiss-born and French-trained vintner, the emphasis at Cuvaison is on traditional French methods. Annual production now runs to twenty thousand cases.

DIAMOND CREEK VINEYARDS / 1500 Diamond Mountain Road, Calistoga, California 94515   Albert Brounstein: (707) 942-6926

OPEN BY APPOINTMENT ONLY. OPEN ONLY EIGHT DAYS EACH SUMMER TO WINE-ORIENTED GROUPS. WT, P.

GR Cabernet Sauvignon; Merlot; Cabernet Franc; Malbec   **R** Silverado Restaurant; Mama Nina's; La Belle Helene   **HM** Wine Country Inn; Calistoga Spa; Silverado Country Club.

DOMAINE CHANDON / P.O. Box 2470, California Drive, Yountville, California 94599   Michaela K. Rodeno: (707) 944-8844

OPEN WEDNESDAY–SUNDAY 11 A.M.–5:30 P.M.; CLOSED NEW YEAR'S DAY, EASTER, CHRISTMAS. WT, RS; D LUNCH AND DINNER. FEE: $1.50 PER FLUTE OF SPARKLING WINE.

GR Pinot Noir; Chardonnay; Pinot Blanc   **R** Domaine Chandon; Chutney Kitchen; La Belle Helene; Mama Nina's; Carriage House; Bon Appetit; Yountville Diner   **HM** Magnolia Hotel; Silverado Country Club; Burgundy House; Chalet Bernensis Inn; Wine Country Inn; Miramonte Hotel; Weber House; Napa Valley Lodge; Harvest Inn.

MISC Domaine Chandon is a wholly owned subsidiary of Moët-Hennessy, a Paris-based holding company that also owns Champagne Moët & Chandon, Mercier and Ruinart, Hennessy Cognac, and Dior Perfumes. Domaine Chandon is unique in that it is the only French-owned winery in California whose parent company assists with the technical aspects of wine making while investing in the California wine industry. The new winery at Yountville will eventually reach a production of 100,000 cases of Sparkling wines per year.

This winery encourages visitors, and small group tours leave from the winery gallery on a frequent basis. Visitors may examine at first hand the production techniques of the *méthode champenoise*, from growing and pressing techniques to primary fermentation, bottle fermentation and aging, riddling, and disgorging and finishing of the Sparkling wine, as they walk through the winery. Wines may be purchased by the glass or by the bottle in Le Salon, where complimentary hors d'oeuvres are also served. The winery boasts a fine French restaurant which offers classic and nouvelle cuisine. A three-course luncheon is available ($8.50), and in the afternoon, coffee, pastries, and wines are available. The dinner menu is a la carte, the chef du cuisine is a native of Champagne, and the ambience is a delight.

CALIFORNIA: THE NAPA VALLEY

FRANCISCAN VINEYARDS / P.O. Box 407, 1178 Galleron Road, Rutherford, California 94573

Justin Meyer or Tim Magill: (707) 963-7111

OPEN DAILY 10 A.M.–6 P.M.; CLOSED NEW YEAR'S DAY, THANKSGIVING, CHRISTMAS. WT, RS.

GR Cabernet Sauvignon; Merlot; Pinot Noir; Gamay; Zinfandel; Chardonnay; White Riesling **R** Mama Nina's; Yountville Restaurant; Domaine Chandon; Magnolia Hotel; Carriage House; Oliver's; Rainbow Bridge; La Belle Helene; Silverado Tavern **HM** Burgundy Inn; Wine Country Inn; Meadowood Suburban Club; Magnolia Hotel **HS** Old Bale Mill; Robert Louis Stevenson Museum.

MISC Although the Franciscan Vineyards are in no way tied into the Franciscan order, their tradition of wine making and hospitality is carried on here. It was the leader of the Franciscan monks, Fra Junípero Serra, who founded the Mission of San Diego in 1769 and who is credited with establishing the first cultivated plantings of grape vines in California. Although there are only two acres of vines planted at this Rutherford vineyard, the Franciscan Vineyards own approximately eight hundred acres under contract in the Sonoma, Lake, and Napa counties.

Visitors are welcome, and may contact Tim Magill for special tours. Visitors are reminded that there are no picnic facilities at this vineyard.

FREEMARK ABBEY / 3022 St. Helena Highway North, St. Helena, California 94574

Charles Carpy: (707) 963-9694

OPEN DAILY 10:30 A.M.–4:45 P.M.; CLOSED EASTER, THANKSGIVING, CHRISTMAS. ONE TOUR DAILY AT 2 P.M. RS, D.

GR Chardonnay; Johannisberg Riesling; Cabernet; Petite Sirah **R** The Abbey Restaurant **HM** Wine Country Inn; El Bonita Motel **HS** The Bale Mill; Bothe State Park; Robert Louis Stevenson Museum.

MISC Freemark Abbey Winery is a modern operation set in a late-nineteenth-century stone cellar. Wines are aged in sixty-gallon French Nevers oak barrels, with a total production of approximately 22,000 cases each year.

HEITZ WINE CELLARS / 500 Taplin Road, St. Helena, California 94558

J. E. Heitz: (707) 963-3542

SALES ROOM OPEN DAILY 11 A.M.–4:30 P.M. WT MONDAY–FRIDAY; RS.

WP Cabernet Sauvignon; Pinot Noir; Grignolino; Barbera Burgundy; Pinot Chardonnay; Chablis; Johannisberg Riesling; Brut and Extra Dry Champagne; Sherry; Port; Angelica.

Heitz Wine Cellars (cont.)

MISC While the Heitz Wine Cellars do not invite touring visitors, the wine tasting room at 436 St. Helena Highway South does offer tasting and retail sales.

INGLENOOK VINEYARDS / P.O. Box 19, Rutherford, California 94573
Vicki Palmer: (707) 963-2616

OPEN DAILY 10 A.M.–5 P.M.; CLOSED NEW YEAR'S DAY, EASTER, THANKSGIVING, CHRISTMAS. TOURS EVERY HOUR ON THE HOUR 10 A.M.–4 P.M. WT, RS; P FOR A FEE.

WP Pinot Noir "Cask"; Cabernet Sauvignon "Cask"; Cabernet Sauvignon; Pinot Noir; Petite Sirah; Charbono; Gamay Beaujolais; Zinfandel; Pinot Chardonnay; White Pinot; Fumé Blanc; Blanc de Noir; Muscat Blanc; Johannisberg Riesling; Gewürztraminer; Sauvignon Blanc; Grey Riesling; Chenin Blanc; Navalle wines; Vintage generics.

MISC These vineyards were founded in 1879 by Finnish sailing captain Gustave Niebaum. Since the Paris Exposition of 1889, where the captain's wines won their first honors, Inglenook has steadily accumulated medals, ribbons, and trophies. Today wines are still aging in old casks brought by Capt. Niebaum from the Black Forests of Germany.

ROBERT KEENAN WINERY / 3660 Spring Mountain Road, St. Helena, California 95674
Joe Cafaro: (707) 963-9177

OPEN WEEKDAYS BY APPOINTMENT ONLY. RS.

GR Chardonnay; Cabernet Sauvignon   **R** La Belle Helene; Domaine Chandon; Miramonte Hotel and Restaurant   **HM** El Bonita Motel; Wine Country Inn.

MISC The first vineyards on this land were planted in the early 1890s by the Conradi Family. Grapes have been crushed here ever since that time. Robert Keenan purchased the land in 1974, and in 1977 he saw the remodeling of the winery and the first crush under his label.

HANNS KORNELL CHAMPAGNE CELLARS / P.O. Box 249, St. Helena, California 94574
Adrian C. Van Dyk, Jr.: (707) 963-9333

CONTACT BY WRITING. OPEN DAILY 10 A.M.–4 P.M.; CLOSED NEW YEAR'S DAY, EASTER, THANKSGIVING, CHRISTMAS. TOURS BY APPOINTMENT ONLY. WT, RS, P.

WP Seven varieties of Champagne: Sehr Trocken; Brut; Extra Dry; Demi Sec; Rosé; Rouge; Muscat Alexandria   **R** Alex's Fine Foods; Freemark Abbey; Square Rigger; Wind Sox Restaurant; Silverado Restaurant

CALIFORNIA: THE NAPA VALLEY

HM Calistoga Spa; Chalet Bernensis Inn; Motel 6 **HS** Old Bale Mill; Napa Valley State Park.

MISC Hanns Kornell came to this country from Germany in 1940. He is the third generation of a family of Champagne makers and he chose to follow his heritage in this country, opening his present business in 1952. In his cellars today there are over 1.5 million bottles of award-winning champagne.

LOWER LAKE WINERY / P.O. Box 950, Lower Lake, California 95457
Thomas Scavone: (707) 994-4069

OPEN BY APPOINTMENT ONLY. RS.

WP Cabernet Sauvignon; Vineyard Specific wines.

MISC The first winery to be reestablished in Lake County since Prohibition, Lower Lake Winery is a small, family-owned-and-operated winery, specializing in the production of Cabernet Sauvignon. The Premiere Vintage will be available in 1980. A 1978 White Cabernet Sauvignon will be available in the summer of 1979.

LOUIS M. MARTINI WINERY / P.O. Box 112, St. Helena, California 94574
Louis P. Martini (707) 963-2736

OPEN DAILY 10 A.M.–4:30 P.M.; CLOSED NEW YEAR'S DAY, EASTER, THANKSGIVING, CHRISTMAS. WT, RS.

WP Cabernet Sauvignon; Gamay; Beaujolais; Johannisberg Riesling; Chenin Blanc; Pinot Noir; Zinfandel **HM** Wine Country Inn; El Bonita Motel; Chalet Bernensis Inn.

MISC Louis Martini was just a young boy when he left his native Pietra Ligure to come to San Francisco. In 1906 the young Louis decided to enter the winemaking field, and he built a small plant in San Francisco. This was the modest forerunner of the Louis Martini Winery in St. Helena today.

J. MATHEWS NAPA VALLEY WINERY / 1711 Main Street, Napa, California 94558
Executive Office, P.O. Box 1042, Newport Beach, California 92663
Kenneth E. Nelson: (714) 642-1234

OPEN BY APPOINTMENT ONLY. RS.

GR Cabernet Sauvignon; Zinfandel; Johannisberg Riesling; Chenin Blanc; Chardonnay **R** The Depot Restaurant; The River House; The Victorian House **HM** Silverado Country Club; Downtown Motel; Wine Country Inn.

J. Mathews Napa Valley Winery (cont.)

MISC Originally built by hand in 1878 by Portuguese stone mason and vintner José Antonio Mateus, this winery has had several owners. The winery's current history centers around Kenneth E. Nelson, who lived for four years with his family in European wine country. Such close association spurred him to the realization of his lifelong dream of owning a winery capable of producing premium wines, and so he assumed his responsibilities as wine maker of the J. Mathews Napa Valley Winery.

MAYACAMAS VINEYARDS / 1155 Lokoya Road, Napa, California 94558
 Robert B. Travers: (707) 224-4030

OPEN BY APPOINTMENT ONLY. RS.

GR Cabernet Sauvignon; Chardonnay.

MISC Originally built in 1889 by John Henry Fisher, this winery has had several owners. The winery was renovated in 1941, and the vineyards replanted to Chardonnay and Cabernet Sauvignon. The Travers Family purchased the winery in 1968.

F. J. MILLER WINERY / 8329 St. Helena Highway, Napa, California 94558
 F. Justin Miller: (707) 963-4252

OPEN BY APPOINTMENT ONLY. WT, RS, P.

WP "Millerway" Effervescent wines.

ROBERT MONDAVI WINERY / 7801 St. Helena Highway, Oakville, California 94562
 Margrit Biever: (707) 963-9611

OPEN DAILY 10 A.M.–5 P.M.; CLOSED MAJOR HOLIDAYS. WT, RS; D IN VINEYARDROOM FOR CATERED PREARRANGED LUNCHEONS OR DINNERS.

GR Cabernet Sauvignon; Chenin Blanc; Chardonnay; Pinot Noir; Sauvignon Blanc; Napa Gamay; Johannisberg Riesling; Muscat Canelli; Merlot **R** La Belle Helene; Hotel Miramonte; The Abbey; Domaine Chandon; Silverado Restaurant and Tavern; Meadowood Country Club **HM** Wine Country Inn; Meadowood Country Club; El Bonita Motel; Hotel Miramonte; Magnolia Hotel; Burgundy Hotel.

MISC For many years Robert Mondavi was general manager of the Charles Krug Winery, which was purchased by Mr. Mondavi's father in 1943. In 1966 Robert Mondavi founded his own winery to provide a future for his children, and it gave him an opportunity to experiment with his own ideas concerning wine making.

CALIFORNIA: THE NAPA VALLEY

NICHELINI VINEYARD / 2349 Lower Chiles Road, St. Helena, California 94574
Jim Nichelini: (707) 963-3357

OPEN SATURDAY AND SUNDAY 10 A.M.–6 P.M.; WEEKDAYS BY APPOINTMENT; CLOSED NEW YEAR'S DAY, THANKSGIVING, CHRISTMAS. WT, RS, P.

GR Cabernet Sauvignon; Semillon; Sauvignon Vert; Zinfandel; Petite Sirah; Carnelian; Chenin Blanc   **HM** El Bonita Motel.

JOSEPH PHELPS VINEYARDS / 200 Taplin Road, P.O. Box 1031, St. Helena, California 94574
Bruce E. Neyers: (707) 963-2745

OPEN MONDAY–SATURDAY 10 A.M.–4 P.M.; CLOSED MAJOR HOLIDAYS. RS.

GR Johannisberg Riesling; Gewürztraminer; Chardonnay; Cabernet Sauvignon; Merlot; Cabernet Franc; Zinfandel; Sirah   **R** Domaine Chandon; Mama Nina's; La Belle Helene; Hotel Miramonte   **HM** Chalet Bernensis Inn; Burgundy House; Magnolia Hotel.

MISC The winery and vineyards were founded in 1972 by Joseph Phelps, who was then joined by German-trained wine maker Walter Schug in developing the winery building and initial 120 acres of vineyard. Each year additional ten-acre plantings have been undertaken, and these will continue until approximately 160 acres are in vines.

RAYMOND VINEYARD AND CELLAR / 849 East Zinfandel Lane, St. Helena, California 94574
(707) 963-3141

OPEN BY APPOINTMENT ONLY. RS.

GR Cabernet Sauvignon; Gamay; Merlot; Zinfandel; Pinot Noir; Chardonnay; Johannisberg Riesling; Chenin Blanc.

MISC The proprietors of the Raymond Vineyard and Cellar note that their family's roots in viticulture and enology in the Napa Valley date back to the 1870s.

RUTHERFORD HILL WINERY / Box 410, St. Helena, California 94574
William P. Jaeger, Jr., or Charles A. Carpy: (707) 963-9694

NOTE: WINERY IS LOCATED AT END OF RUTHERFORD HILL ROAD, OFF SILVERADO TRAIL, RUTHERFORD. OPEN SECOND SATURDAY OF EACH MONTH, WITH TASTING; OTHERWISE, BY APPOINTMENT ONLY.

WP Johannisberg Riesling; Gewürztraminer; Pinot Noir; Pinot Noir Blanc; Zinfandel; Cabernet Sauvignon; Merlot; Chardonnay   **R** The French Laundry   **HM** Wine Country Inn; Napa Valley Lodge.

Rutherford Hill Winery (cont.)

MISC This winery was originally built in 1972 and was owned by a subsidiary of the Pillsbury Company until it was bought by the present owners in 1976.

RUTHERFORD VINTERS / 1673 St. Helena Highway South, Rutherford, California 94573   Bernard L. Skoda: (707) 963-4117

OPEN DAILY 10 A.M.–4:30 P.M.; CLOSED NEW YEAR'S DAY, EASTER, THANKSGIVING, CHRISTMAS. TOURS BY APPOINTMENT. WT, RS, P.

GR Johannisberg Riesling; Cabernet Sauvignon   **WP** Johannisberg Riesling; Cabernet Sauvignon; Pinot Noir; Muscat of Alexandria.

MISC Although only recently bonded in September 1977, the Rutherford Vintners soon plan annual production of 15,000 to 20,000 cases. The capacity is 83,000 gallons.

V. SATTUI WINERY / White Lane, St. Helena, California 94574
Daryl Sattui: (707) 963-7774

OPEN DAILY 9:30 A.M.–6 P.M. WT, RS, P.

WP Zinfandel; Cabernet Sauvignon; Burgundy; Johannisberg Riesling, Rosé of Cabernet Sauvignon.

MISC The V. Sattui Winery features an extensive gift shop with more than 150 varieties of cheese, as well as homemade pâtés, chocolates, home-smoked cheddars, homemade cheese cakes, breads, and assorted gourmet foods. The wines are sold exclusively at the winery, not in stores or restaurants.

SCHRAMSBERG VINEYARDS / Schramsberg Road, Calistoga, California 94515   Jack L. Davies: (707) 942-4558

OPEN MONDAY–SATURDAY BY APPOINTMENT ONLY. RS.

GR Chardonnay; Pinot Noir   **R** Silverado Tavern; La Belle Helene; Hotel Miramonte; The Abbey   **HM** Harvest Inn; El Bonita; Wine Country Inn   **HS** Old Bale Mill; Silverado Museum; Robert Louis Stevenson State Park.

MISC This winery, founded in 1862, was described by R. L. Stevenson in his *Silverado Squatters*. The vineyards were reopened in 1965 by Jack Davies, a specialist in bottle-fermented champagne.

CALIFORNIA: THE NAPA VALLEY 65

SILVER OAK CELLARS / 915 Oakville Crossroads, Oakville, California 94562 or P. O. Box 414, Oakville, California 94562
Justin Meyer: (707) 944-8866

OPEN BY APPOINTMENT ONLY. RS AT FRANCISCAN WINERY (SEE LISTING).

GR Cabernet Sauvignon.

MISC The Silver Oak Cellars and the Franciscan Winery are under the same ownership, and share some personnel and services. Of special note is the retail sale of Silver Oak Cabernet at Franciscan.

SMITH-MADRONE VINEYARDS / 4022 Spring Mountain Road, St. Helena, California 95674   Stuart or Charles Smith: (707) 963-2283

OPEN BY APPOINTMENT ONLY. RS, P.

GR Chardonnay; Cabernet Sauvignon; Pinot Noir; Johannisberg Riesling.

MISC This mountain vineyard features a hand-built winery with underground cellars and a sod roof. Vintner Stuart Smith has taught viticulture and enology at local community colleges.

SPRING MOUNTAIN VINEYARDS / 2805 Spring Mountain Road, St. Helena, California 94574
Michael Robbins or Julie Moretton: (707) 963-4341

OPEN MONDAY–FRIDAY 2:30 P.M. BY APPOINTMENT ONLY.

GR Cabernet Sauvignon; Chardonnay; Sauvignon Blanc; Pinot Noir   **R** La Belle Helene; Miramonte; Domaine Chandon; Silverado Restaurant and Tavern   **HM** Wine Country Inn; Chalet Bernensis Inn.

STAG'S LEAP WINE CELLARS / 5766 Silverado Trail, Napa, California 94558   (707) 944-2020

OPEN MONDAY–FRIDAY 8 A.M.–4 P.M. RS; P LIMITED.

GR Cabernet Sauvignon; Merlot   **R** Mama Nina's; Domaine Chandon.

MISC This young vineyard, planted in 1970, was established after a five-year search for a set of characteristics most nearly ideal for California Claret.

STAG'S LEAP WINERY / 6150 Silverado Trail, Napa, California 94558
Carl Doumani: (707) 944-2792

OPEN BY APPOINTMENT ONLY.

GR Cabernet; Petite Sirah; Chenin Blanc; Merlot; Pinot Noir.

MISC This winery was founded in 1888. It is presently being restored by its new owners, and the winery began operations in 1970.

ST. CLEMENT VINEYARDS / 2867 St. Helena Highway North, St. Helena, California 94574   Dr. William J. Casey: (707) 963-7221

OPEN BY APPOINTMENT ONLY. RS (CASE LOTS).

GR Pinot Chardonnay; Cabernet Sauvignon   **R** La Belle Helene; Hotel Miramonte; French Laundry; Magnolia Hotel.

MISC The St. Clement Vineyards feature a historic Victorian house constructed in 1876, with a winery built into the hand-cut stone cellar. The current owners, Dr. William Casey and his wife, Alexandra, limit their annual production to under eight thousand cases.

STERLING VINEYARDS / 1111 Dunaweal Lane, Calistoga, California 94515   Valerie Presten: (707) 942-5151

OPEN DAILY MAY 1–OCTOBER 31 10:30 A.M.–4:30P.M.; OPEN WEDNESDAY–SUNDAY NOVEMBER 1–APRIL 30 10:30 A.M.–4:30 P.M.; CLOSED NEW YEAR'S DAY, EASTER, THANKSGIVING, CHRISTMAS. APPOINTMENTS NECESSARY FOR GROUPS OF TWELVE OR MORE. TRAMWAY FEE: $2.50. WT, RS; D (CATERED ONLY).

GR Cabernet Sauvignon; Merlot; Pinot Noir; Zinfandel; Chardonnay; Sauvignon Blanc; Gewürztraminer; Chenin Blanc.

MISC The visit includes an aerial tramway ride to the hilltop winery, self-guided tour of the winery, and wine tasting. Sterling Vineyards also accommodates prearranged groups of 12–150 for guided tours, and catered luncheons and dinners.

STONEGATE WINERY / 1183 Dunaweal Lane, Calistoga, California 94515   David Spaulding: (707) 946-6500

OPEN DAILY 10 A.M.–4 P.M. BY APPOINTMENT ONLY; CLOSED MAJOR HOLIDAYS. RS, P.

GR Chardonnay; Sauvignon Blanc; Pinot Noir; Merlot; Cabernet Sauvignon   **R** La Belle Helene; Domaine Chandon; Magnolia Hotel; Silverado Restaurant and Tavern; Mama Nina's   **HM** Calistoga Spa; Wine

CALIFORNIA: THE NAPA VALLEY 67

Country Inn   **HS** Napa Valley Wine Library; Robert Louis Stevenson Museum; "Old Faithful" Geyser.

MISC Originally planted in 1971 by James and Barbara Spaulding, the vineyards, as well as the winery, are now under the management of their son David. He is assisted by Michael Fallow, a trained viticulturist.

STONY HILL VINEYARD / P.O. Box 308, St. Helena, California 94574
Mrs. McCrea: (707) 963-2636

OPEN BY APPOINTMENT ONLY. RS.

GR Chardonnay; White Riesling; Gewürztraminer; Semillon.

MISC The owners comment: "We are glad to show Stony Hill to really interested people, but we have no separate touring staff, and therefore have to limit our time spent on touring. Our vineyards now cover approximately thirty acres. The area is hilly, so that some of the fields are terraced, and the cultivation methods are therefore somewhat old-fashioned. Our output is very small, never more than three thousand cases, but much sought after."

SUTTER HOME WINERY / 277 St. Helena Highway South, St. Helena, California 94574   Roger J. Trinchero: (707) 963-3104

OPEN DAILY 9 A.M.–5 P.M.; CLOSED MAJOR HOLIDAYS. WT 10 A.M.–4:30 P.M.; RS.

WP Zinfandel; White Zinfandel; Moscato Amabile   **R** The Wine Barrel; Miramonte Hotel; La Belle Helene; Silverado Restaurant and Tavern
HM Chalet Bernensis Inn; El Bonita; Wine Country Inn   **HS** Robert Louis Stevenson Museum, Old Bale Mill.

MISC This winery, originally constructed in 1874, was purchased by the Trinchero Family in 1946. The Trincheros represent a long family heritage of wine making dating back six generations to the Asti region in northern Italy.

TREFETHEN VINEYARDS / 1160 Oak Knoll Avenue, Napa, California 94558   Janet Trefethen: (707) 255-7700

OPEN BY APPOINTMENT ONLY. RS.

GR Chardonnay; White Riesling; Gewürztraminer; Pinot Noir; Merlot; Zinfandel; Cabernet Sauvignon   **WP** Chardonnay; White Riesling; Cabernet Sauvignon; Pinot Noir   **R** Domaine Chandon; Mama Nina's; Bon Appetit; Silverado Restaurant and Tavern; The French Laundry; Miramonte Inn   **HM** Wine Country Inn; Magnolia Hotel; Chalet Bernensis Inn; Meadowood Country Club   **HS** Old Bale Mill.

Trefethen Vineyards (cont.)

MISC This is one of the oldest wooden tongue-and-groove wineries in the Napa Valley, with a capacity of 200,000 gallons. The winery and six hundred surrounding acres were purchased in 1968 by the Trefethen Family. The vineyard was planted and developed under the guidance of Tony Baldini.

TULOCAY WINERY / 1426 Coombsville Road, Napa, California 94558
W. C. Cadman: (707) 255-4699

OPEN BY APPOINTMENT ONLY. RS.

WP Cabernet Sauvignon; Pinot Noir; Zinfandel; Chardonnay; Pinot Blanc **HS** Tulocay Cemetery.

MISC All wines are crushed, fermented, aged, and bottled on the premises. The first release, a Cabernet Sauvignon, was in November 1978.

VILLA MOUNT EDEN / Oakville Crossroads, Oakville, California 94562
Nils Venge: (707) 944-8431

OPEN DAILY BY APPOINTMENT ONLY. WT, RS, P.

GR Cabernet Sauvignon; Pinot Chardonnay; Pinot Noir; Chenin Blanc; Gewürztraminer; Napa Gamay.

YVERDON VINEYARDS / 3787 Spring Mountain Road, St. Helena, California 94574 (707) 963-4270

OPEN MONDAY–FRIDAY 8 A.M.–3 P.M. RS.

GR Cabernet Sauvignon; Gamay; Johannisberg Riesling; Chenin Blanc; Gewürztraminer.

MISC These vineyards were started in 1970 by Fred and Russ Aves.

California: The Sonoma Region

Few areas in California or elsewhere can match Sonoma County for the beauty of its pastoral scenery. Driving through this superb wine region can be a memorable experience in itself. And likewise the tasting, because in the 1940s Sonoma was the scene of a great revival in wine growing in California, and some of the state's finest vintages come from this district.

The early Franciscan fathers founded their northernmost winery in 1823 at the Sonoma Mission, where they planted the first wine grapes in Northern California. The great wine history of the area continued in the 1850s with the efforts of two colorful rivals, Count Agoston Haraszthy and General Mariano Vallejo. Count Haraszthy, a Hungarian immigrant, was commissioned in 1856 by the California legislature to study the prospects for wine growing in the region. In that year he bought and named the vineyards of Buena Vista, and he later imported to Sonoma 100,000 of the finest European cuttings. He also sent his sons back to Europe to study wine making in the great French districts. Haraszthy cultivated in competition with his famous neighbor, General Vallejo, who had taken over and planted the vineyards of the old Sonoma Mission. Their rivalry became a friendly one when Haraszthy's two sons married Vallejo's daughters in a double ceremony at the Mission. Despite Haraszthy's varied fortunes in growing, his was a tremendous effort, and his 100,000 cuttings, distributed throughout the valley, became the basis for experimentation in this first great wine-growing region in the state. As you travel north to the Russian River area and all the way to Mendocino, the vineyards you visit are all likely to contain descendants of Haraszthy's early imports. Sonoma celebrates its wine history, and the post-Prohibition revival in growing, with a festival in the town of Sonoma during the last week of September.

If you wish to enter the valley through some spectacular country, you can take coastal Route 1 and stop at the remarkable Point Reyes National Park, with its redwood preserve and magnificent seashore; then proceed east on Route 12, entering wine country at the junction of Route 101. Or you can take 101 north directly from the Golden Gate Bridge, turn east on Route 37 at Ignacio, and north on Route 121 to Route 12, which leads into Sonoma. If you wish to venture farther north to the vineyards around Mendocino (again, spectacular seacoast scenery), Route 101 will take you there too.

A symbol of the revival in wine production in this area was the revitalization of Haraszthy's old Buena Vista Vineyards in the 1940s, and large-scale operations in the valley really date from the past three decades. A few wineries have functioned more or less continuously since the great days of the nineteenth century, and have consistently won awards for the quality of their products. Picnic areas, both at vineyards and at recreational facilities, are numerous throughout the valley. This area has one of the most gentle and beautiful inland landscapes in America—warm in the valleys and crisp on the hills. And you'll find links with an international (Russian and Spanish) colonial past. It is worthwhile, even in the case of public facilities, to inquire beforehand about availabilities and reservations, since this area is really too good to miss.

ALEXANDER VALLEY VINEYARDS / 8644 Highway 128, Healdsburg, California 95448   Hank H. Wetzel III: (707) 433-6293
OPEN DAILY 8 A.M.–5 P.M.; CLOSED MAJOR HOLIDAYS. WT, RS.

GR Chardonnay; Johannisberg Riesling; Gewürztraminer; Chenin Blanc; Cabernet Sauvignon; Merlot; Pinot Noir; Zinfandel   **R** Belvedere Restaurant; Rex Restaurant   **HM** Fairview Motel   **HS** Several nineteenth-century buildings at the vineyard.

CALIFORNIA: THE SONOMA REGION

MISC The Wetzel Family purchased this property in 1963, and plantings for their vineyards took place over the next decade. The name of the vineyards comes from Cyrus Alexander, who homesteaded this land in 1842.

BUENA VISTA WINERY / 18000 Old Winery Road, Sonoma, California 95476
Rose Murphy: (707) 938-1266

OPEN DAILY 10 A.M.–5 P.M. CLOSED NEW YEAR'S DAY, THANKSGIVING, CHRISTMAS. WT, RS, P.

GR Cabernet Sauvignon; Pinot Noir; Gamay Beaujolais; Johannisberg Riesling; Gewürztraminer; Pinot Chardonnay **R** Gino's; Depot Hotel Restaurant; Au Relais; La Casa; Mariano's **HS** Sonoma Mission; Vallejo's Home.

MISC The Buena Vista vineyards were planted in the Sonoma Valley—the old "Valley of the Moon" of Jack London—in 1832, to produce wine for the Mission San Francisco Solano de Sonoma. Twenty years later the vineyards were acquired by Count Agoston Haraszthy, a titled vintner from the court of Emperor Ferdinand of Hungary. Count Haraszthy had fled to America after his participation in the Hapsburg Revolution, and he concluded his long flight in the Sonoma Valley, where he introduced European varietals and laid the foundation for the California wine industry. This extraordinary wine maker prospered in the New World, and before his death in 1869 he became sheriff and assemblyman for San Diego, commissioner for the State of California, and the first director of the San Francisco Mint. At Buena Vista his two sons married the daughters of General Mariano Vallejo in a double wedding at the old Mission. The winery and the vineyards, fallen into disuse, were restored by Frank Bartholomew, then head of United Press International, in 1943. Today Buena Vista is owned by Young's Market Company of Los Angeles.

CAMBIASO VINEYARDS / 1141 Grant Avenue, Healdsburg, California 95448
Somchai Likitprakong: (707) 433-5508

OPEN MONDAY–SATURDAY 10 A.M.–4 P.M.; CLOSED MAJOR HOLIDAYS. RS.

GR Cabernet Sauvignon; Petite Sirah; Zinfandel. **R** House of Sonoma.

MISC These vineyards were planted from selected cuttings of fine grape varietals in the 1920s, and for a few years the mountain-grown grapes were sold to individuals who made their own wine in the basements of their homes in San Francisco. In 1934 Giovanni Cambiaso and his wife Maria came from Italy to establish this winery, which has been in the Cambiaso Family ever since.

CHÂTEAU SONOMA WINERY / Wohler and River Roads, Forestville,
California 95436   Ira Gourvitz: (415) 343-4134

OPEN BY WRITTEN APPOINTMENT ONLY.

GR Chenin Blanc; French Colombard; Semillon.

MISC Château Sonoma was founded in 1977 by the S.S. Pierce Company.

CHÂTEAU ST. JEAN / Box 293, 8555 Sonoma Highway, Kenwood,
California 95452   Richard L. Arrowood: (707) 833-4134

OPEN DAILY 10A.M.–4:30 P.M. WT, RS; P HIGHLY RECOMMENDED.

GR Johannisberg Riesling; Gewürztraminer; Chardonnay; Sauvignon
Blanc **R** The Depot Inn; Belvedere; Black Forest Inn; Au Relais; Villa
HM Hillside Motel; Los Robles Lodge; Sonoma Mission Inn;
TraveLodge **HS** Hood Mansion; Sugarloaf State Park; Jack London
Estate and Museum.

MISC Although the Château St. Jean is a relatively new addition to the
Sonoma palette of wines, the high quality of the château's wines was apparent at the 1976 Sonoma County Harvest Fair, where the château took seven of the seventy-seven medals awarded. Owners Ken Sheffield and Robert and Edward Merzoian, experienced grape growers, credit their young wine maker, Richard Arrowood, for much of the château's success. Arrowood did graduate work in enology at California State University in Fresno. He has worked with Korbel Champagne Cellars (see listing), where he was assistant to Champagne Master Allan Hemphill, now president of Château St. Jean; he has been with Italian Swiss Colony (see listing) and the Sonoma Vineyards (see listing). In 1977 he served as president of the Sonoma Valley Vintners' Association.
 Complementing the château's high standards is the care taken to ensure a visitor's enjoyment. Arrowood sums up this philosophy by saying, "We want people to come to the château, taste a glass or two of wine, see the villa and stroll around the gardens; to enjoy themselves. We hope many will bring a picnic and make a real day out of buying a little wine for their cellar."

CLOS DU BOIS / 36 Mill Street, Healdsburg, California 95448
              Frank M. Woods: (415) 456-7310

OPEN BY APPOINTMENT ONLY.

WP Gewürztraminer; Johannisberg Riesling; Cabernet Sauvignon; Pinot
Noir; Chardonnay.

MISC Owner Frank Woods has extensive vineyard holdings in Dry Creek and Alexander Valley. When his new premium varietal vineyards matured in the early '70s, he began to produce wines exclusively from his own vineyards. His

CALIFORNIA: THE SONOMA REGION

limited bottlings of 100 percent varietal wines have already won critical acclaim, and have been awarded medals in California wine competitions.

CRESTA BLANCA WINERY / 2399 North State Street, Ukiah, California 95482 (707) 462-0565

OPEN DAILY 9 A.M.–5 P.M.; CLOSED NEW YEAR'S DAY, EASTER, THANKSGIVING, CHRISTMAS. WT, RS, P.

GR Red and White varietals.

MISC *Time* magazine credited the Cresta Blanca Mendocino Zinfandel as "the best buy in American red wine," but praise for Cresta Blanca wines dates back to the nineteenth century. It was in 1883 that California vintner Charles A. Wetmore returned from France with cuttings from Château d'Yquem and Château Margaux, and grafted them onto American rootstock. In 1889 Wetmore returned to France to enter his first vintages in the Paris Exposition. Out of some seventeen thousand wine entries, Cresta Blanca placed first, receiving the *grand prix*. His friend and source for the cuttings, the Marquis de Lur-Saluces, exclaimed, "The daughter has excelled the mother."

DACH VINEYARDS / 3700 Highway 128, Philo, California 95466
John Dach: (707) 895-3245

DACH RANCH (Fruit Stand) / 9200 Highway 128, Philo, California 95466
(707) 895-3173

OPEN BY APPOINTMENT ONLY. RS.

GR Pinot Chardonnay; Gewürztraminer; Pinot Noir; fifty-seven varieties of apples also grown here, as well as vegetables, including sweet corn (all organic) **HS** Fort Bragg; Skunk Steam Train.

MISC Owner and wine maker John Dach, formerly of the Novitiate at Los Gatos (see listing: East Bay–Livermore, California), came to Philo with his wife, Sandi, in 1972. Adding their own vineyards to the already existent apple orchard, John and Sandi plan to sell their own hard cider, apple wine, and wines from their winery at the fruit stand they operate.

DRY CREEK VINEYARD / 3770 Lambert Bridge Road, Healdsburg, California 95448 (707) 433-1000

OPEN DAILY 10:30 A.M.–4:30 P.M.; CLOSED MAJOR HOLIDAYS. WT, RS, P.

Dry Creek Vineyard (cont.)

GR Chenin Blanc; Sauvignon Blanc; Chardonnay; Cabernet Sauvignon
R Mama Nina's   **HM** Los Robles Lodge.

MISC Founded in 1972 by David Stare, the Dry Creek Vineyard wines won three gold medals at the 1977 Los Angeles and Sonoma county fairs.

EDMEADES VINEYARDS / 550 California State Highway 128, Philo, California 95466   (707) 895-3232

OPEN DAILY 11 A.M.–6 P.M. WT 25¢ PER TASTE; RS, P.

GR Cabernet Sauvignon; Pinot Chardonnay; Gewürztraminer; French Colombard   **WP** Cabernet Sauvignon; Zinfandel; Pinot Chardonnay; Gewürztraminer; French Colombard; Pinot Noir; Mendocino Rain wine; Queen Anne's Lace; Mendocino Whale wine; and two Apple wines
R Ledford House; St. Orres Inn; McCallum House; Elk Cove Inn; Heritage House; Bear Wallow Lodge; Little River Inn; Palace Bar and Grill; Leonard's Bistro   **HM** Mendocino Hotel; Little River Inn; McCallum House; Heritage House; Harbor House; Bear Wallow Lodge; Palace Hotel; Philo Motel
HS Fort Ross; historic town of Mendocino.

MISC This second-generation vineyard and winery now produces 24,000 gallons of wine annually.

FETZER VINEYARDS / 1150 Bel Arbres Road, Redwood Valley, California 95470   Patti Box: (707) 485-8998 or 485-8802

OPEN MONDAY–FRIDAY 10 A.M.–2 P.M. BY APPOINTMENT ONLY. WT, RS, P.

GR Cabernet Sauvignon; Zinfandel; Petite Sirah   **R** The Broiler Steak House; Club Calpella Restaurant   **HM** Ridgewood Park Motel; LuAnn Motel   **HS** Mendocino County Museum.

MISC The Fetzer Family moved to this historic 750-acre ranch in the 1950s. Their winery was established in 1968. Today three generations of the Fetzer Family are working the vineyards and the winery. Fetzer Vineyards also has a tasting room and retail wine sales room in the village of Hopland, California, fifteen miles south of Ukiah on U.S. 101. It's open from 9 A.M. to 5 P.M. seven days a week.

FIELDBROOK VALLEY WINERY / Route 1, Box 314, Fieldbrook, California 95521   Bob Hodgson: (707) 839-4140

OPEN BY APPOINTMENT ONLY.

CALIFORNIA: THE SONOMA REGION

GR Riesling.

MISC This new winery, established in 1976, has an annual yield of less than two hundred gallons.

FIELD STONE WINERY / 10075 Highway 128, Healdsburg, California 95448 Sally Hepper: (707) 433-7266

OPEN DAILY 9 A.M.–5 P.M.; CLOSED MAJOR HOLIDAYS. RESERVATIONS NECESSARY FOR GROUP VISITS. RS, P.

GR Chenin Blanc; Gewürztraminer; Johannisberg Riesling; Cabernet Sauvignon; Petite Sirah **R** Wine Country Restaurant; Healdsburg House; Souverain Restaurant; Catelli's The Rex **HM** Sheraton Tropicana; Los Robles Lodge; L & M Motel.

MISC Owner Wallace Johnson, former mayor of Berkeley, California, planted his vineyards twelve years ago. This winery, built in 1977, produces 100 percent varietal wines.

FOPPIANO VINEYARDS / 12707 Old Redwood Highway, Healdsburg, California 95448 Louis M. Foppiano: (707) 433-1937

OPEN DAILY 10 A.M.–4:30 P.M.; CLOSED MAJOR HOLIDAYS. WT, RS.

GR Cabernet Sauvignon; Sauvignon Blanc; Chardonnay; Petite Sirah; Zinfandel; Pinot Noir; French Colombard **R** Giorgio's; House of Sonoma; Catelli's; Wine Country Restaurant **HS** Fort Ross; Luther Burbank Gardens.

MISC In the words of the current owners: "Great grandfather John arrived from Italy in 1864 to seek his fortune in the goldfields. Instead, he found his life's work in the vineyards near Healdsburg. His oldest son, Louis, took over the winery in 1910. In those early days wooden casks of wine went by horse-drawn wagons to be sold in old San Francisco, where customers brought their own jugs for filling. Today the third and fourth generations of the Foppiano Family are carrying on the family tradition."

GEYSER PEAK WINERY / P.O. Box 198, Geyserville, California 95441 Millie Howie: (707) 433-5349

OPEN DAILY 10 A.M.–5 P.M.; CLOSED MAJOR HOLIDAYS. WT, RS, P.

GR Pinot Chardonnay; Zinfandel; Pinot Noir; Cabernet Sauvignon; Sauvignon Blanc; Johannisberg Riesling **HS** Geysers; geothermal wells.

Geyser Peak Winery (cont.)

MISC Geyser Peak Winery was founded in 1880 by Augustus Quitzow. In 1971 new buildings and equipment were added, and the winery now produces a complete line of vintage-dated wines under the Geyser Peak label, as well as jug wines under the Summit label. In 1976 Geyser Peak introduced "Wine-in-a-box" and in 1977 they produced a new, lightly spritzed Grand Blanc. There are two walking trails at Geyser Peak, and during the summer, by reservation, there is a moonlight walk and tasting.

GRAND CRU VINEYARDS / 1 Vintage Lane, Glen Ellen, California 95442
Mail to: P.O. Drawer B, Glen Ellen, California 95442
Allen B. Ferrera: (707) 996-8100

OPEN SATURDAY AND SUNDAY 10 A.M.–5 P.M.; WEEKDAYS AND GROUPS BY APPOINTMENT ONLY; CLOSED NEW YEAR'S DAY, THANKSGIVING, CHRISTMAS. WT 25¢ PER TASTE; RS, P.

GR Zinfandel **R** Bunny's Country Kitchen; The Golden Bear Lodge; The Grapes Cafe **HM** The London Lodge Motel **HS** Jack London State Park and Museum.

MISC This modern and mechanized winery, founded in 1971 by Robert Magnani and Allen Ferrera, is situated among vineyards originally planted in 1890.

GRAND PACIFIC VINEYARD COMPANY / 134 Paul Drive, #8, San Rafael, California 94903
(415) 479-9463

OPEN DAILY 8 A.M.–5 P.M. WT, RS.

WP Cabernet Sauvignon; Merlot; Chardonnay; White Riesling; Muscat Canelli.

MISC This small winery was founded in 1975 by Richard Berry Dye, a wine maker who also operates the California Steam Beer Brewing Company next door to Grand Pacific. Mr. Dye is currently planting Marin County vineyards.

GUNDLACH-BUNDSCHU WINERY / 3775 Thornsberry Road, Vineburg, California 95487
John Merritt: (707) 938-5277

OPEN FRIDAY–SUNDAY NOON–5 P.M.; CLOSED MAJOR HOLIDAYS. WT, RS, P.

WP Zinfandel; Cabernet Sauvignon; Johannisberg Riesling; Kleinburger.

CALIFORNIA: THE SONOMA REGION

MISC Although the current Gundlach-Bundschu Winery was founded in 1970, it retains a strong sense of historic tradition. The walls were built in 1858 when the original Gundlach-Bundschu Winery opened a business which was to become world famous. Those nineteenth-century wines were distributed throughout the United States, winning awards at home as well as in Paris and Guatemala. However, this great business was ill-fated and met disaster in the San Francisco earthquake and fire. As with so many wineries, Prohibition brought production to a virtual standstill. But on Halloween, 1970, three young members of the Bundschu Family got together over a glass of homemade wine and decided to restore the Gundlach-Bundschu reputation and business. They ran into some opposition to using the name Gundlach-Bundschu from Towle Bundschu, who felt very strongly about the history of the name and what it had stood for. But when he tasted the first new wines, he gladly endorsed it.

HACIENDA WINE CELLARS / 1000 Vineyard Lane, Sonoma, California 95476
Steven W. MacRostie: (707) 938-3220

OPEN DAILY BY APPOINTMENT; CLOSED NEW YEAR'S DAY, THANKSGIVING, CHRISTMAS. WT, RS, P.

GR Cabernet Sauvignon; Chardonnay; Johannisberg Riesling; Pinot Noir; Merlot; Gewürztraminer; Zinfandel; Sylvaner   **R** Au Relais; Bon Voyage; Capri   **HM** El Pueblo Motel   **HS** Sonoma Mission; Vallejo Home.

MISC Hacienda Wine Cellars was established in 1973 by Frank Bartholomew (see listing under Buena Vista Winery). The winery also owns Buena Vista Vineyards, established in 1862 by Agoston Haraszthy. The winery was purchased in 1977 by A. Crawford Cooley. Steve MacRostie has been the wine maker here since 1974.

HANZELL VINEYARDS / 18596 Lomita Avenue, Sonoma, California 95476
Bob Sessions: (707) 996-3860

OPEN BY APPOINTMENT ONLY. RS WHEN AVAILABLE.

GR Chardonnay; Pinot Noir.

HOP KILN WINERY AT GRIFFIN VINEYARDS / 6050 Westside Road, Healdsburg, California 95448
John R. Warne or Martin Griffin, M.D.: (707) 433-6491

TASTING ROOM OPEN WEEKENDS 10 A.M.–5 P.M., OR BY APPOINTMENT. WT, RS, P.

Hop Kiln Winery at Griffin Vineyards (cont.)

GR Zinfandel; Petite Sirah; Early Burgundy; Chardonnay; French Colombard; Johannisberg Riesling; Gamay Beaujolais; Gewürztraminer.

MISC The Hop Kiln Winery is a state historical landmark, and features a tasting room for its premium varietal wines. It serves as a museum of the agricultural history of Sonoma County. Griffin Vineyards is the first "historic agricultural district" so designated by a county in California. The vineyards were first planted in 1880.

HUSCH VINEYARDS / 4900 Star Route, Philo, California 95466
(707) 895-3216

OPEN DAILY 10 A.M.–5 P.M. WT, RS.

GR Pinot Noir; Chardonnay; Gewürztraminer; Johannisberg Riesling; Cabernet Sauvignon.

MISC In the sales room at Husch Vineyards you will also find watercolors by Gretchen Husch.

ITALIAN SWISS COLONY / P.O. Box 1, Asti, California 95413
Write to: Hospitality Manager   (707) 894-2280

OPEN DAILY 9 A.M.–5 P.M.; TOURS EVERY HOUR; CLOSED NEW YEAR'S DAY, EASTER, THANKSGIVING, CHRISTMAS. WT, RS, P.

GR Numerous varietals   **R** Hoffman House; Catelli's   **HM** La Grande Motel   **HS** Vallejo's Home; Homes of Luther Burbank and Jack London; geothermal geysers.

MISC Established in 1881 by Andrea Sbarboro with a group of Italian and Swiss immigrants on sixteen hundred acres, this winery's first wines were produced in 1886. The Colony soon won worldwide acclaim for its gold-medal–winning vintages. The Colony is now a California historical landmark and contains the largest collection of redwood cooperage in the world.

JOHNSON'S ALEXANDER VALLEY / 8333 Highway 128, Healdsburg, California 95448   Jay or Tom Johnson: (707) 433-2319

OPEN DAILY 10 A.M.–5 P.M.; CLOSED NEW YEAR'S DAY, THANKSGIVING, CHRISTMAS. WT, RS, P; D (CATERED PRIVATE GROUPS).

GR Chardonnay; Chenin Blanc; Gewürztraminer; Zinfandel; Pinot Noir; Cabernet Sauvignon; Johannisberg Riesling   **R** Cloverdale Inn;

CALIFORNIA: THE SONOMA REGION

Cricklewood; Souverain; House of Sonoma; Tamaulipico; Healdsburg House **HM** Fairview Motel; Holiday Inn; TraveLodge **HS** Luther Burbank Gardens; Fort Ross.

MISC This small, family-owned-and-operated vineyard and winery also features occasional Sunday concerts on an old theater pipe organ.

KENWOOD VINEYARDS / P.O. Box 447, Kenwood, California 95452
Martin Lee: (707) 833-5891

OPEN DAILY 9 A.M.–5 P.M. WT, RS, P.

GR Johannisberg Riesling; Zinfandel **R** Bunny's Country Kitchen **HS** Jack London Ranch; Sugarloaf Ridge State Park.

MISC This winery was built in 1906 and produced bulk wines until 1970, when it was purchased by six wine enthusiasts from San Francisco. Since then the winery produces small quantities of premium wines, which are now sold in most major cities. They are best known for their high quality Cabernet Sauvignons and Zinfandels.

F. KORBEL AND BROTHERS / 13250 River Road, Guerneville, California 95446
Gary B. Heck: (707) 887-2294

OPEN DAILY 9 A.M.–5:30 P.M.; CLOSED NEW YEAR'S DAY, EASTER, CHRISTMAS. WT, RS, P.

GR Pinot Chardonnay; Gewürztraminer; Johannisberg Riesling; Pinot Blanc; Cabernet Sauvignon; Pinot Noir **R** Northwood Restaurant; Hexagon House; Buck's; Burdon's **HM** Northwood Lodge; Hexagon House; Murphy's Resort **HS** Fort Ross; Armstrong Woods State Park.

MISC Francis, Joseph, and Anton Korbel came to this country in the mid-1850s. They purchased extensive timberland in the lower Russian River area of Sonoma and planted European grapevines. Since their first crush in 1881 their business has prospered. This is one of the few wineries in America producing Champagne by the traditional bottle-fermented method, and Korbel Champagnes are known throughout the world.

LAMBERT BRIDGE / 4085 West Dry Creek Road, Healdsburg, California 95448
Ed Samperton: (707) 433-5855

OPEN BY APPOINTMENT ONLY. RS.

Lambert Bridge (cont.)

WP Cabernet Sauvignon; Chardonnay; Merlot   **HM** Los Robles Lodge; Fairview.

MISC The Lamberts limit their annual production to ten thousand cases.

LANDMARK VINEYARDS / 9150 Los Amigos Road, Windsor, California 95492   William or Michele Mabry: (707) 838-9466

OPEN WEDNESDAY AND FRIDAY 1 P.M.–5 P.M.; SATURDAY AND SUNDAY 10 A.M.–5 P.M.; AND BY APPOINTMENT; CLOSED MAJOR HOLIDAYS. RS, P.

GR Cabernet Sauvignon; Pinot Noir; Chardonnay; Gewürztraminer; White Riesling   **R** Cricklewood; Larkfield; Belvedere Restaurant   **HM** Los Robles Lodge; Holiday Inn.

MISC Opening in 1972, Landmark now has vineyards in the Sonoma and Alexander valleys, and in Windsor. This small winery features an original ranch house dating back to 1850 which the current owners hope to make into a small wine museum.

LYTTON SPRINGS / 650 Lytton Springs Road, Healdsburg, California 95448 Write to: Dee Sindt, 15101 Keswick Street, Van Nuys, California 91405   (707) 433-7721

OPEN BY APPOINTMENT ONLY. RS.

WP Zinfandel.

MISC This winery, opened in 1977, is located in the Valley Vista Vineyards which have been growing Zinfandel grapes for over eighty years. One of the owners, Dee Sindt, is also the publisher of *Wine World* magazine.

MARK WEST VINEYARDS AND WINERY / 7000 Trenton–Healdsburg Road, Forestville, California 95436   Joan Ellis: (707) 544-4813

OPEN BY APPOINTMENT ONLY. RS, P.

GR Gewürztraminer; Johannisberg Riesling; Chardonnay; Pinot Noir   **R** River's End; L'Omelette; Mark West Lodge; La Provence   **HM** Hexagon House; Los Robles Lodge; River Wood Resort.

MISC All wines are estate grown and bottled at this new family winery.

CALIFORNIA: THE SONOMA REGION

MARTINI AND PRATI WINES, INC. / 2191 Laguna Road, Santa Rosa, California 95401　　　　　　　　　Frank Vannucci: (707) 823-2404

OPEN MONDAY–FRIDAY 9 A.M.–4 P.M. WT, RS.

GR Zinfandel; Gamay Beaujolais; Pinot Noir; Petite Sirah; Merlot; Chenin Blanc; Grey Riesling; French Colombard　　**R** Lena's　　**HM** Los Robles Holiday Inn　　**HS** Luther Burbank Gardens.

MISC This winery was founded at the turn of the century by Raphaele Martini and his sons. It was later briefly owned by Hiram Walker, but in 1950 Elmo Martini and Enrico Prati, formerly with Italian Swiss Colony (see listing), repurchased the family winery. The winery now features two labels: Martini & Prati and Fountaingrove.

MILANO WINERY / 14594 South Highway 101, Hopland, California 95449　　　　　　　　　　　　　James A Milone: (707) 744-1360

OPEN BY APPOINTMENT ONLY. WT, RS, P.

WP Cabernet Sauvignon; Petite Sirah; Zinfandel; Chenin Blanc; Cabernet.

MISC Using grapes grown by several third-generation vintners from this area, the wine makers of the Milano Winery say, "The right grape from the right place makes the most distinctive wine."

MILL CREEK VINEYARDS AND WINERY / 1401 Westside Road, Healdsburg, California 95448　　　　　　Bill Kreck: (707) 433-5098

OPEN BY APPOINTMENT ONLY. WT, P (FACILITIES PLANNED); RS.

GR Pinot Chardonnay; Cabernet Sauvignon; Pinot Noir; Merlot　　**R** Los Robles Lodge Restaurant　　**HM** Los Robles Lodge.

MISC Although this new winery produced and bottled its first wines in 1976, it has already been awarded two gold medals, two silver medals, and a bronze, at the Sonoma County Harvest Fair.

NAVARRO VINEYARDS / 5751 Highway 128, Philo, California 95466　　　　　　　　　　　　　　Deborah Cahn: (707) 895-3686

OPEN BY APPOINTMENT ONLY. RS, P.

GR Pinot Noir; Gewürztraminer; Riesling　　**R** Ledford House; Cafe Beaujolais　　**HM** Mendocino Hotel; Heritage House; Harbor House.

PARDUCCI WINE CELLARS / 501 Parducci Road, Ukiah, California 95482
John A. Parducci: (707) 462-3828

OPEN DAILY IN SUMMER 9 A.M.–6 P.M.; 9 A.M.–5 P.M., WINTER; CLOSED MAJOR HOLIDAYS. TOURS EVERY HOUR ON THE HOUR, EXCEPT AT THE BEGINNING OF THE WORK DAY, THE END OF THE WORK DAY, AND NOON. WT, RS, P.

GR Cabernet Sauvignon; Chardonnay; Gamay Beaujolais; Pinot Noir; Chenin Blanc; Petite Sirah; French Colombard; Sylvaner Riesling; Flora; Semillon; Carignane **R** The Green Barn; Leonard's Bistro; Manor Inn; Broiler Steak House **HM** Holiday Lodge; Lu-Ann Motel **HS** United States Observatory; Willits Museum.

MISC Parducci Cellars was founded in 1931 by Adolph Parducci. Second- and third-generation Parduccis now run these vineyards and cellars. Utilizing grapes from three different Parducci vineyards, these family cellars emphasize traditional wine-making processes.

J. PEDRONCELLI WINERY / 1220 Canyon Road, Geyserville, California 95441
James Pedroncelli: (707) 857-3619

OPEN DAILY 10 A.M.–5 P.M.; CLOSED NEW YEAR'S DAY, EASTER, THANKSGIVING, CHRISTMAS. WT, RS.

GR Gewürztraminer; Chardonnay; Pinot Noir; Zinfandel; Johannisberg Riesling; Gamay; Cabernet Sauvignon **R** Catelli's The Rex; Hoffman House; Wine Country Inn.

MISC This winery has been in the Pedroncelli Family since 1927. Production emphasis is on the classical varietals, with fermentation in temperature-controlled stainless-steel tanks, and aging in both wood and stainless steel.

POPE VALLEY WINERY / 6613 Pope Valley Road, Pope Valley, California 94567
(707) 965-2192

OPEN DAILY 11 A.M.–6 P.M. CALL IN ADVANCE. WT, RS, P.

WP Chenin Blanc; Semillon; White Riesling; Sauvignon Blanc; Pinot Chardonnay; Pinot Noir; Petite Sirah; Zinfandel; Cabernet Sauvignon.

MISC This seventy-year-old winery was purchased by the Devitt Family in 1972. Their 1975 Pinot Noir won a bronze medal at the 1977 Los Angeles County Fair, and their 1976 Muscat of Alexandria won a gold medal in that competition.

CALIFORNIA: THE SONOMA REGION

PRESTON WINERY / 9282 West Dry Creek Road, Healdsburg, California 95448   (707) 433-4748
NOT OPEN TO THE PUBLIC AT THIS TIME.

GR Sauvignon Blanc; Zinfandel; Chenin Blanc; Cabernet Sauvignon   **R** Catelli's The Rex; Mama Nina's.

A. RAFANELLI / 4685 West Dry Creek Road, Healdsburg, California 95448   Americo or Mary Rafanelli: (707) 433-1385
OPEN BY APPOINTMENT ONLY. RS BY CASE.

GR Zinfandel; Gamay Beaujolais; French Colombard   **R** Healdsburg House; House of Sonoma   **HM** L & M Motel.

MISC This winery, which limits annual production to two thousand cases of estate-bottled Zinfandel and Gamay Beaujolais, was founded in 1974.

SAUSAL WINERY / 7370 Highway 128, Healdsburg, California 95448
    David Demostene: (707) 433-2285
NOT OPEN TO THE PUBLIC.

GR Zinfandel; Cabernet Sauvignon; Napa Gamay; Carignane; Pinot Noir.

MISC Although this small family-owned winery, founded in 1973, is not open to the public, Sausal does send information about their new releases to people on their mailing list.

SEBASTIANI VINEYARDS / 389 Fourth Street East, Sonoma, California 95476   (707) 938-5532
OPEN DAILY 10 A.M.–5 P.M.; CLOSED NEW YEAR'S DAY, EASTER, THANKSGIVING, CHRISTMAS. WT, RS.

GR Barbera; Cabernet Sauvignon; Chardonnay; Green Hungarian; Pinot Noir; White Riesling; Zinfandel   **R** Au Relais; Capri; Gino's; La Casa; El Dorado Hotel; Swiss Hotel; Giarritta's Sugar and Spice; Bon Voyage Cafe   **HM** El Pueblo Motel   **HS** Mission San Francisco de Solano; Blue Wing Inn; Vallejo's Home.

MISC This wine cellar is a state landmark. The winery operation here was started in 1903 by Samuele Sebastiani, and it has been completely owned and operated by the Sebastiani Family ever since. A wide assortment of wines is available in the tasting room, and occasionally one of the proprietor's re-

Sebastiani Vineyards (cont.)

serve selections is opened for tasting. In the last fifteen years these vineyards have won 320 top awards in international wine and spirit competitions. Tours begin every twenty minutes.

SIMI WINERY / P.O. Box 946, 16275 Healdsburg Avenue, Healdsburg, California 95448   Patricia Vadon: (707) 433-6981

OPEN DAILY 10 A.M.–5 P.M.; CLOSED MAJOR HOLIDAYS. TOURS AT 11 A.M., 1 P.M., AND 3 P.M. WT, RS, P.

WP Gewürztraminer; Chenin Blanc; Johannisberg Riesling; Chardonnay; Cabernet Sauvignon; Pinot Noir; Zinfandel; Gamay Beaujolais; Rosé of Cabernet Sauvignon.

MISC Giuseppe Simi came to California in the years of the Gold Rush and for some thirty years he tried his luck at mining and farming. When he and his brother Pietro first saw the green hills of Alexander Valley and the nearby town of Healdsburg, they were struck by the resemblance to their native village. Here, they felt, was the perfect place for a winery. By 1876 they were ready to make their move toward a lifetime ambition. They created their own winery, a rustic building of rough hand-hewn stone, and carefully selected the finest grapes available. Today Simi Alexander Valley wines are enjoyed throughout the nation. Andre Tchelistcheff, master enologist, acts as consultant at Simi. These vineyards publish an occasional and informative newsletter, which is available upon request without charge.

SONOMA COUNTY COOPERATIVE WINERY / P.O. Box 36, Windsor, California 95492   Cecil de Loach: (707) 838-6649

OPEN BY APPOINTMENT ONLY.

GR Zinfandel; Petite Sirah; French Colombard; Cabernet Sauvignon; other varietals.

SONOMA VINEYARDS / Windsor, California 95492
Tasting rooms: 72 Main Street, Tiburon; and 2191 Union Street, San Francisco; and at the winery.   (707) 433-6511

OPEN DAILY 10 A.M.–4 P.M.; CLOSED MAJOR HOLIDAYS. WT, RS, P.

GR Chardonnay; Pinot Noir; Cabernet Sauvignon; Zinfandel; Johannisberg Riesling; French Colombard; Chenin Blanc.

CALIFORNIA: THE SONOMA REGION

MISC Founded in 1970, this publicly held winery produces five estate-bottled wines, vintage-dated varietals, and "fermented-in-this-bottle" Champagne.

SOTOYOME WINERY / 641 Limerick Lane, Healdsburg, California 95448
William Chaikin: (707) 433-2001

OPEN BY APPOINTMENT ONLY. WT, RS.

WP Chardonnay; Cabernet Sauvignon; Petite Sirah; Zinfandel
GR Zinfandel; Petite Sirah   **R** Healdsburg House; Catelli's; Los Robles Lodge   **HS** Fort Ross; numerous houses, churches, and other nineteenth-century buildings.

MISC The Sotoyome Winery, named after the Mexican Land Grant on which it is situated, commenced production with the vintage of 1974. This is a small winery with an annual production of four thousand cases.

SOUVERAIN CELLARS / P.O. Box 528, Geyserville, California 95441
Ed King: (707) 433-6918

OPEN DAILY 10 A.M.–5 P.M.; CLOSED NEW YEAR'S DAY, EASTER, THANKSGIVING, CHRISTMAS. TOURS DAILY FROM 10 A.M.–4 P.M. APPOINTMENT NECESSARY FOR GROUPS OF TWENTY OR MORE. WT, RS, D.

WP Cabernet Sauvignon; Pinot Noir; Petite Sirah; Gamay Beaujolais; Zinfandel; Pinot Noir Rosé; Chardonnay; Johannisberg Riesling; Colombard Blanc; Grey Riesling; Dry Chenin Blanc; Burgundy; Chablis   **R** Souverain Restaurant at vineyard; Hoffman House   **HM** Wine Country Inn; Magnolia Inn; Los Robles Lodge; Sandman Inn; Heritage House; McCallum House **HS** Armstrong Redwood Park.

MISC Souverain is owned by a limited partnership of 220 north coast growers. Souverain has won several gold medals for its wines, which are 100 percent varietal and 100 percent vintage. The vineyards feature a terrace restaurant serving luncheon from 11 A.M. daily, and dinner from 5 P.M.–9 P.M. daily except Monday and Tuesday.

TRENTADUE WINERY / 19170 Redwood Highway, Geyserville, California 95441
Leo Trentadue: (707) 433-3104

OPEN DAILY 10 A.M.–5 P.M.; CLOSED NEW YEAR'S DAY, EASTER, THANKSGIVING, CHRISTMAS. WT, RS; P (PLANNED).

GR Cabernet Sauvignon; Gamay; Merlot; Early Burgundy; Petite Sirah; Carignane; Zinfandel; Chenin Blanc; Semillon; French Colombard; Pinot Chardonnay; Johannisberg Riesling   **R** Catelli's the Rex; Souverain

Trentadue Winery (cont.)

Winery Restaurant   **HM** Fairview Motel; Flamingo Hotel; Los Robles Inn; Oaks Motel   **HS** Asti Vineyards; Luther Burbank House.

MISC Established in 1969, this family-owned-and-operated winery features a large gift shop next to the wine-tasting room. The shop offers personalized labels.

VALLEY OF THE MOON VINEYARDS / 777 Madrone Road, Glen Ellen, California 95442   Harry Parducci: (707) 996-6941

OPEN DAILY, EXCEPT THURSDAY, 10 A.M.–5 P.M.; CLOSED NEW YEAR'S DAY AND CHRISTMAS. WT, RS, P.

GR Zinfandel; French Colombard; Semillon   **R** La Casa; Bar Capri; Swiss Hotel; Sonoma Grove   **HM** London Lodge Motel; Sonoma Mission Inn; El Pueblo Motel; Sonoma Hotel   **HS** Depot Museum; Jack London State Park; Sonoma Valley Historical Society.

MISC The Valley of the Moon Winery has been owned and operated by the Parducci Family since 1941. However, the colorful history of these vineyards goes back over a century. Originally a portion of the Agua Caliente Rancho, granted by the Mexican government to Lazara Pena, the land was purchased by General M. G. Vallejo, and later, 640 acres were given to his children's music teacher in exchange for their piano lessons. In 1851 Joseph Hooker took over this portion of the ranch and planted a vineyard, using Indian or Chinese labor. Eli T. Sheppard, former American consul to Tienstsin, and later an advisor in international law to the Japanese emperor, bought the property in 1883 and named it The Madrone Vineyards. He added French vines to the vineyards, and was written up in several books of that time as one of the growers whose names are almost as well known as the wines of Sonoma themselves. Because of poor health, he sold the vineyards to Senator George Hearst and retired to San Francisco in 1888. When Enrico Parducci purchased the Madrone vineyards, the winery had fallen into disuse. However, the Parduccis were able to start production in 1942, and under the supervision of Otto Toschi, the Valley of the Moon Winery has been in operation for over thirty years.

WEIBEL CHAMPAGNE VINEYARDS / Home winery: 1250 Sanford Avenue, Mission San Jose, California 94538
Second tasting room: 7051 North State Street, Redwood Valley, California
                    Marlene F. Weibel: (415) 656-2340 and (707) 485-0321

BOTH LOCATIONS OPEN DAILY 10 A.M.–5 P.M.; CLOSED CHRISTMAS. APPOINTMENT NECESSARY FOR LARGE GROUPS. WT, RS, P.

CALIFORNIA: THE SONOMA REGION

WP Premium Red, White, and Rosé wines; Solera Flor Sherries; Champagnes **R** Fremont Inn Restaurant; Spin-A-Yarn Restaurant; Mission Delicatessen **HM** Fremont Motor Inn **HS** Mission San Jose de Guadalupe.

MISC This winery was founded in 1869 by Leland Stanford, governor of California, United States senator, and founder of Stanford University. The Weibel Family, originally from Switzerland, acquired the winery in 1945. Three generations of this family now operate the winery, and their success is reflected by over 650 awards for excellence in the United States and abroad.

WILLOW CREEK VINEYARD / 1904 Pickett Road, McKinleyville, California 95521   Dean Williams (707) 839-3373

OPEN BY APPOINTMENT ONLY. WT, RS.

GR Merlot; Cabernet Sauvignon; Gewürztraminer; Zinfandel; Chardonnay **R** Big Foot Golf Club; Lazia's **HM** Eureka Inn; Caison House Motel **HS** Hoopa Indian Reservation; Redwood Park.

MISC Willow Creek is situated in the mountains, surrounded by state and federal forests and the gigantic redwoods. The winery was bonded in 1976. (McKinleyville is situated in Humboldt County, far to the north of Sonoma.)

ZD WINERY / P.O. Box 900, Sonoma, California 95476
          Gino Zepponi, Norman de Leuze: (707) 938-0750 or 539-9137

TOURS BY APPOINTMENT ONLY. WT, RS.

WP Pinot Noir; Chardonnay; Gewürztraminer; Cabernet Sauvignon **HM** El Pueblo; Sonoma Mission Inn; Old Sonoma Hotel **HS** Sonoma Mission; Vallejo's Home; Jack London Home.

MISC The Zepponi and de Leuze families have been making wine for generations. The ZD winery was founded in 1969 when the current owners, previously engineering associates, discovered a mutual interest in wines. After studying enology at the University of California at Davis, they set out to start their winery with a grape crusher and press available from the Zepponi Family. Their philosophy: "Small wineries produce the exceptional wines which require the care of time-consuming methods and devotion to quality which is possible only through personal attention."

The Pacific Northwest (Oregon, Washington, Idaho, and British Columbia)

To a history of increasing recognition for fine wines, wine makers of the Pacific Northwest are adding an era of innovation and expansion in the planting of hybrid grapes. The prospects in these areas are promising. People now compare the isolated wine districts of Oregon to the early years of the Napa Valley, and most vineyards in the area are small and carefully run. They are already noted for the success of their efforts by wine lovers throughout America.

Oregon's wine industry is centered around two locations: the Roseburg area and the Willamette and Tualatin valleys. In these districts the proper amount of rain, a good soil, and long summer days combine to make a hearty environment for Pinot Noir, White Riesling, and Chardonnay, as well as for fine European-American hybrid stocks. Oregon vintners, with long experience in the production of fruit wines, are increasingly eager to compare their fine premium wines against the best California can offer. Greater numbers of wine enthusiasts find this pride in Oregon wines quite justifiable.

Washington vintners note that their state is the third largest grape producer in America. Many acres of Concord grapes are being replaced by *V. vinifera* grape plantings. Many Washington vineyards are irrigated, but more modest ones share the good fortune of the location of Washington wine growing—the Yakima Valley is on the same latitude as the great French wine districts. This leads to firm hopes for competition with the European products some day, especially given the long northerly summer days and the fine grape soil of the region. (Yakima is on Route 82, which cuts off from Route 90 between Seattle and Spokane roughly halfway through the state.)

Since the places of interest to wine lovers are scattered throughout the Northwest, it is advisable to be in no hurry. The wineries listed here will guide you along the intervening frontier country and provide you with some memorable wine tasting as well.

Oregon

AMITY VINEYARDS / Route 1, Box 348-B, Amity, Oregon 97101
Myron Redford: (503) 835-2362

OPEN IN SUMMER TUESDAY–SUNDAY 11 A.M.–6 P.M.; OPEN REST OF YEAR SATURDAY AND SUNDAY NOON–5 P.M.; CLOSED DECEMBER 31 TO JANUARY 31 AND AT OTHER TIMES BY ADVANCE NOTICE. WT, RS, P.

GR Pinot Noir; White Riesling; Chardonnay; Gewürztraminer; Semillon; Pinot Gris   **R** Nick's Italian Cafe; La Michiuana; The Bayou; Amity Cafe; S.A.G.E. Cafe   **HM** Travelers 7 Motel.

MISC This vineyard offers two annual wine, food, and music festivals. Nouveau wines are released at the Winter Solstice Wine Festival, and White wines are released at the Summer Solstice Wine Festival. There is an admission charge.

BJELLAND VINEYARDS / Route 4, Box 931, Roseburg, Oregon 97470
Paul and Mary Bjelland: (503) 679-6950

OPEN DAILY IN WINTER 11 A.M.–5 P.M.; SUMMER 8 A.M.–5 P.M.; CLOSED HOLIDAYS AND OCCASIONALLY ON TUESDAYS. WT, RS, P.

GR *Vinifera*   **R** Windmill; Single Tree.

MISC The Bjelland Cabernet Sauvignon and Wild Blackberry wines have been awarded medallions at the Oregon State Fair.

THE PACIFIC NORTHWEST

CHARLES COURY VINEYARDS / Box 372, Forest Grove, Oregon 97116
Charles Coury: (503) 357-7602

OPEN WEDNESDAY–SUNDAY 1 P.M.–5 P.M. WT, RS, P.

GR *Vinifera*   **R** Red Baron; Copperstone Restaurant   **HM** Forest Grove Motel   **HS** Tillamook Burn; Vernonia Railroad; Scotch Church

MISC Wine maker Charles Coury earned his master's degree in viticulture and enology at the University of California at Davis. At the winery is a one-hundred-year-old house in the basement of which wines were formerly made.

ELK COVE VINEYARDS / Route 3, Box 23, Gaston, Oregon 97119
Joe or Patricia Campbell: (503) 985-7760

OPEN BY APPOINTMENT ONLY. WT (WHEN ANNOUNCED); RS, P.

GR Pinot Noir; Chardonnay; Gewürztraminer; White Riesling.

MISC This small family-owned-and-operated vineyard is named after the magnificent Roosevelt Elk, which can be seen during the Oregon winter season.

EYRIE VINEYARDS / P.O. Box 204, Dundee, Oregon 97115
David or Diana Lett: (503) 864-2410 or 472-6315

OPEN BY APPOINTMENT ONLY. RS.

GR Pinot Noir; Chardonnay; Riesling; Pinot Gris; Pinot Meunier; Muscat Ottonel   **R** Nick's Italian Cafe; S.A.G.E. Cafe   **HM** Safari Motel; Travelers 7 Motel.

MISC The Eyrie Vineyards are the first *vinifera* vineyards in the Willamette Valley since Prohibition. Planted in 1965, winery production began in 1970. Although the winery offers no public tastings, these are offered semiannually to those on the Eyrie Winery mailing list, for which visitors may apply for inclusion.

FORGERON VINEYARDS / 89697 Sheffler Road, Elmira, Oregon 97437
George L. Smith: (503) 935-3530

OPEN WEEKENDS BY APPOINTMENT ONLY. WT, RS, P.

GR Pinot Noir; Beaujolais; Riesling; Pinot Chardonnay; Gewürztraminer
HM Valley River Inn.

Forgeron Vineyards (cont.)

MISC In 1970 George and Linda Smith planted twenty-five acres of vines, and their own winery was licensed in 1977. Now the vineyard also has a testing plot from Oregon State University where new experimental vine varieties are being evaluated.

HILLCREST VINEYARD / 240 Vineyard Lane, Roseburg, Oregon 97470
(503) 673-3709

OPEN DAILY 10 A.M.–5 P.M. WT, RS.

WP White Riesling; Gewürztraminer; Fumé Blanc; Chenin Blanc; Rosé of Cabernet; Cabernet Sauvignon; Zinfandel; Mellow Red.

MISC The Hillcrest Vineyard represents the fruition of Richard Sommer's dream to discover a new viticultural region for the production of quality wines in Oregon. A fourteen-acre hillside farm was purchased in 1961 and planted that spring. Full production was achieved by 1966. A new winery was constructed in 1975, and eventual capacity will be fifty-thousand gallons, with grapes coming from different Douglas County vineyards.

HONEY HOUSE WINERY / 26202 Fawver Road, Veneta, Oregon 97487
Patricia or Robert Saxton: (503) 935-2008

OPEN BY APPOINTMENT ONLY. RS, P.

WP Mead, Dry and Sweet; Blackberry; Rhubarb; Apple   **HS** Pioneer Museum, University of Oregon Museum of Art and Science.

MISC This winery, founded in 1976, takes its name from the old farm honey house, where honey was extracted from the family apiary. Today extracted honeys are made into fine meads. These may be purchased under the Willamette Valley Homestead label.

JONICOLE VINEYARDS / Route 1, Box 1118, Roseburg, Oregon 97470
Jon Marker: (503) 679-6435

OPEN WEEKENDS NOON–5 P.M. WT, RS.

GR Cabernet Sauvignon; Chardonnay; White Riesling   **R** Homestead Specialty Shop   **HM** Garden Villa.

MISC The vineyard was planted in 1968 and the winery was built in 1973.

THE PACIFIC NORTHWEST   93

KNUDSON ERATH WINERY / Route 1, Box 368, Dundee, Oregon 97115
(503) 538-3318

OPEN WEEKENDS AND SOME HOLIDAYS 11 A.M.–5 P.M. WT, RS.

GR Pinot Noir; White Riesling; Chardonnay; Merlot; Gewürztraminer.

MISC C. Calvert Knudson and Richard C. Erath pooled their viticultural resources in 1975 to form the present winery, with an annual capacity of 35,000 gallons.

NEHALEM BAY WINERY / Route 1, Box 440, Nehalem, Oregon 97131
Patrick O. McCoy: (503) 368-5300

OPEN DAILY 10 A.M.–5 P.M. WT, RS, P.

WP Pinot Noir; Cabernet Sauvignon; Zinfandel; White Riesling; Grey Riesling; Chardonnay; Cherry; Loganberry; Blackberry; Plum; Peach; Apricot; Currant; Cranperé   **R** Manzanita Inn; Crab Broiler
HS Tillamook County Pioneer Museum; Indian Treasure Mountain (Neahkah-nie).

MISC This winery, founded in 1972 by Pat McCoy, was built in a converted cheese factory. The attractive tasting room is furnished entirely from local sources.

OAK KNOLL WINERY / Route 4, Box 185-B, Hillsboro, Oregon 97123
Ron Vuylsteke: (503) 648-8198

OPEN THURSDAY, FRIDAY, AND SUNDAY 2 P.M.–6 P.M.; SATURDAY 11 A.M.–6 P.M. WT, RS.

WP Loganberry; Blackberry; Raspberry; Boysenberry; Rhubarb; Gooseberry; Strawberry; Red Currant; Plum; Pinot Noir; Chardonnay; Zinfandel; Rosé; White Riesling; Cabernet Sauvignon; Niagara   **R** Jake's Crayfish Restaurant; Benson Hotel.

MISC Although the Hillsboro area had many wineries around the turn of the century, these vanished completely with Prohibition. Oak Knoll was one of the first wineries to begin production again, in 1970. The owners claim that their winery is referred to locally as "The Château Lafite of Fruit and Berry wines."

PONZI VINEYARDS / Route 1, Box 842, Beaverton, Oregon 97005
Richard Ponzi: (503) 628-1227

OPEN BY APPOINTMENT ONLY. WT, RS, P.

Ponzi Vineyards (cont.)

GR Pinot Noir; Chardonnay; Riesling; Pinot Gris   **R** L'Auberge; Belinda's; Couch Street Fish House; Genoa; Jake's Crayfish; Imbrey Farmstead
HM Portland Hilton; Sheraton; Thunderbird; Greenwood Inn   **HS** Oregon Museum of Science and Industry.

MISC This family-operated vineyard and winery, bonded in 1974, has already won honors in county, regional, and national competitions.

SOKOL BLOSSER WINERY / Blanchard Lane, Dundee, Oregon 97115
William or Susan Blosser: (503) 864-3342 or 864-2307

OPEN WEEKENDS. WT, RS, P; D (CATERED ONLY).

GR Chardonnay; Pinot Noir; White Riesling; Müller-Thurgau; Sauvignon Blanc; Merlot   **R** Nick's Italian Cafe   **HM** Safari Motel.

MISC This new winery, the vineyards of which were planted in 1971, began with an annual capacity of twenty thousand cases. The tasting room features a gift shop which sells fresh fruit and nuts in season.

TUALATIN VINEYARDS, INC. / Route 1, Box 339, Forest Grove, Oregon 97116   (503) 357-5005

OPEN SATURDAY AND SUNDAY 1 P.M.–5 P.M.; CLOSED MAJOR HOLIDAYS. WT, RS, P.

WP White Riesling; Muscat of Alexandria; Gewürztraminer; Petite Sirah; Pinot Noir   **HM** Holiday Motel   **HS** Portland Museum and Zoo.

MISC Wine maker William L. Fuller notes that Tualatin was established in 1973. It now has some seventy acres under vines and a fifty-thousand-gallon winery.

Washington

ASSOCIATED VINTNERS, INC. / Winery: 4368 150th Avenue, N.E., Redmond, Washington   Vineyard: Sunnyside, California
L. S. Woodburne: (206) 883-1146

CALL FOR APPOINTMENT AND DIRECTIONS. WT LIMITED; RS.

THE PACIFIC NORTHWEST

GR Gewürztraminer; Riesling; Chardonnay; Semillon; Pinot Noir; Cabernet **HM** Holiday Inn; Thunderbird Motel **HS** Pioneer Square; Seattle Center and Science Center

MISC This thirty-acre vineyard was planted in 1962. Its first commercial vintage was released in 1967.

CHÂTEAU STE. MICHELLE / P.O. Box 1976, 14111 N.E. 135th, Woodinville, Washington 98072   Bob Betz: (206) 485-9721

OPEN DAILY 10 A.M.–4:30 P.M.; CLOSED NEW YEAR'S DAY, THANKSGIVING, CHRISTMAS EVE, CHRISTMAS. WT, RS, P.

GR Johannisberg Riesling; Semillon; Chenin Blanc; Grenache; Cabernet Sauvignon; Merlot; Gewürztraminer; Chardonnay; Sauvignon Blanc; Pinot Noir; Muscat **R** Gerard's; Relais de Lyon; Schroeder's.

MISC Château Ste. Michelle is located on eighty-seven acres of arboretumlike ground. Buildings from the turn-of-the-century estate are still standing, among trout ponds, lakes, and streams. In 1976 a modern winery was built and opened to the public for daily cellar tours and wine tasting.

HINZERLING VINEYARDS / 1520 Sheridan Avenue, Prosser, Washington 99350   Mike, Jerry, or Dee Wallace: (509) 786-2163

OPEN BY APPOINTMENT ONLY. WT, RS.

GR Gewürztraminer; White Riesling; Chardonnay; Cabernet Sauvignon; Merlot **R** Gasperetti's; Balcom's.

MISC The vineyard was planted in 1972 and the first crush was in 1976.

LEONETTI CELLARS / 1321 School Avenue, Walla Walla, Washington 99362   Gary Figgins: (509) 525-1428

OPEN BY APPOINTMENT ONLY. WT, RS.

GR Cabernet Sauvignon; Merlot; White Riesling **R** Black Angus Restaurant **HM** Black Angus Motel **HS** Whitman Mission National Historical Site and Museum

MISC Inspired by grandfather Frank Leonetti, the current owners of the Leonetti Cellars have established Cabernet Sauvignon and White Riesling on the Leonetti estate and Merlot at the winery site. With an average annual rainfall of seventeen inches, these vines do not require irrigation.

OCEAN SHORES WINERY / 884 North Ocean Shores Boulevard, Ocean Shores, Washington 98567

OPEN DAILY 9 A.M.–6 P.M.; CLOSED DURING WINTER MONTHS AND MAJOR HOLIDAYS. WT, RS.

WP All Fruit and Berry wines including: Raspberry; Rhubarb; Red Currant; Pear   **R** Ocean Shores Inn Restaurant   **HM** Gray Gull Motel; Ocean Shores Inn; Gitchie-Goomie Motel.

MISC This winery is a branch of the Puyallup Winery (see next listing).

PUYALLUP VALLEY WINERY / 121 23rd Street S.E., Puyallup, Washington 98371   (206) 848-4573

OPEN MONDAY–SATURDAY 9 A.M.–6 P.M.; SUNDAY NOON–6 P.M.; CLOSED MAJOR HOLIDAYS. WT, RS.

WP Rhubarb; Raspberry; Red Currant; Blackberry; Tart Cherry; Pear; Apricot; Black Cherry; Gooseberry; Marionberry   **R** Antone's; Puyallup Elks Club; Iron Gate   **HM** Nendel's Motel   **HS** Frontier Museum; Meeker Mansion.

MISC The current owners contend that the Puyallup Valley Winery began as a practical joke when, at a meeting of Washington State Berry Growers, someone suggested that a winery would be a good way to promote berry sales. The winery now produces some twelve flavors and plans an additional four or five.

MANFRED J. VIERTHALER WINERY / 17136 Highway 410 East, Sumner, Washington 98390   (206) 863-1633

OPEN TUESDAY–SUNDAY NOON–6 P.M. WT, RS, P.

GR White Riesling; Müller-Thurgau   **R** Bavarian Chalet Restaurant   **HM** Bavarian Chalet Motel   **HS** Meeker Mansion.

MISC This winery, built and designed by owner Vierthaler, sold its first bottle of wine in January 1977. Vierthaler was invited to join the exclusive *Rheingauer Weinconvent* in 1970. The winery features an attractive tasting room in the Bavarian style.

Idaho

STE. CHAPELLE VINEYARDS / P.O. Box 729, Emmett, Idaho 83617
Bob Schoenwald: (208) 365-5162
OPEN TUESDAY–SUNDAY 1 P.M.–5 P.M. WT, RS, P.

GR Riesling; Chardonnay; Sauvignon Blanc; Pinot Noir; Cabernet Sauvignon; Merlot; Gewürztraminer.

MISC The vineyards at Ste. Chapelle are seven years old, and the winery was founded in 1976. A new winery is under construction in the Sunny Slope area near Caldwell, Idaho. The cellar room is completed and the tasting and touring rooms are scheduled to open in June 1979. The tasting room in Emmett will remain open until then.

British Columbia

ANDRE'S WINES (B.C.) LTD. / 2120 Vintner Street, Port Moody, British Columbia, Canada (604) 937-3411
OPEN DAILY IN SUMMER NOON–3 P.M.; ALSO TUESDAY–THURSDAY 7 P.M.–8 P.M. WT.

WP Red, White, and Rosé Table wines; Champagnes; Baby Duck; Chantes; Perles; Aperitifs **R** Mulvaney's; Old Spaghetti Factory; Puccini's; Three Greenhorns; Medieval Inn; Hy's Encore **HM** Sheraton Villa; Bayshore Inn; Georgia Hotel **HS** Port Moody Museum; Port Moody Historical Society; Historical Churches.

MISC One of six modern wineries owned by Andrew Peller (see listing, Great Lakes Region), the Port Moody winery is a modern facility with equipment for the pressing, aging, and bottling of both Still wines and Champagnes.

CALONA WINES LTD. / 1125 Richter Street, Kelowna, British Columbia, Canada (604) 762-3332
OPEN JUNE–SEPTEMBER MONDAY–FRIDAY 10 A.M.–3 P.M.; CLOSED HOLIDAYS. WT, RS.

Calona Wines Ltd. (cont.)

WP Table, Sparkling, Fruit, and Dessert wines **R** Capri Hotel Dining Room; Keg and Cleaver; La Bussola; Talos Greek Restaurant **HM** Capri Motel; Royal Ann Hotel; Caravel Motor Inn; Sandman Inn **HS** Father Pandosy Mission; Kelowna Centennial Museum.

MISC This winery is located in the heart of the Okanagan Valley, which, with its excellent trout and kokanee fishing, has become a popular vacation area. Calona Wines was founded in 1932 by local businessmen, including W. A. C. Bennett, who later became premier of this province. It is now the largest winery in British Columbia, with a storage capacity of four million gallons.

CASABELLO WINES, LTD. / 2210 Main Street, Penticton, British Columbia, Canada A. R. Wareham: (604) 492-0621

OPEN MONDAY–FRIDAY 10:30 A.M.–4:30 P.M. WT, RS.

WP Varietals; Champagnes; Crackling, Sparkling, Dessert, and Table wines **HS** Penticton Museum.

MISC Casabello has been producing wines by traditional European methods since 1966. Tours are conducted on the half hour and the wine store and tasting room are open evenings.

STE. MICHELLE WINES / 15050 54-A Avenue, Surrey, British Columbia, Canada (604) 576-6741

OPEN BY APPOINTMENT ONLY. WT, RS.

WP Table and Dessert wines; Varietals; Sparkling Cider **HM** Metro Vancouver **HS** Fort Langley; Peach Arch Park.

MISC Ste. Michelle Wines originated in 1932 as Growers Wine Company, which produced Loganberry wines from the berries growing on Vancouver Island. In 1926, with the development of the first vineyards in the Okanagan Valley, grape wines were produced. The new modern Ste. Michelle winery opened in 1978 and is located in Surrey, British Columbia. It has a tank storage capacity of four million gallons. A winery retail store, wine tasting room, and winery tours are features of the new premises.

The Ohio River Valley (Ohio, Indiana, and Kentucky)

The Ohio River Valley is rich in wine lore, with a wine history dating to the early nineteenth century. It was in the years when Pittsburgh was a western outpost of the early states that the first experiments in viticulture were being undertaken along the Ohio. The earliest production of wine in what is now the state of Ohio began with Moravian missionaries to the Indians and with French settlers who had brought French vines to the area around present-day Marietta. By the middle of the nineteenth century a wine industry with common grapes and growing conditions had been established along the Ohio River from Marietta to the tip of Indiana, where the Wabash meets the Ohio in the west. The Ohio Valley wine area is represented today primarily by wines grown in Ohio and Indiana, though Kentucky, too, knew a time of wine growing.

As the Ohio River marks out the border between Ohio and West Virginia and then Kentucky, it cuts the sharp, wooded hills of the area, providing much dramatic scenery. The steamboats that carried freight and passengers between Pittsburgh, Cincinnati, and the Mississippi in the first half of the last century would have passed landings shipping a distinguished early-American Champagne from the north shore of the Ohio. As the hills begin to roll more gently where the Ohio borders Indiana, two historically important wine regions are also located in the eastern and western tips of that state. The stretch of the Ohio where Indiana and Ohio meet has been known as the "Rhineland" of America, and both states have seen exciting moments in the cultivation of grapes.

After early French and Moravian wine growing, which was on a small scale, the story of Ohio wines along the river really begins with a Cincinnati lawyer, Nicholas Longworth. In the 1820s he abandoned his law practice in favor of the cultivation of grapes, and especially of a variety he had discovered under cultivation in nearby Georgetown. This local grape was the Catawba, and it was grown so successfully around Cincinnati that by the 1850s Longworth's Still Catawba was thought in London to be finer than any Rhine Hock, and his Sparkling Catawba was America's first Champagne. The poet Longfellow was inspired to write an "Ode to Catawba Wine." By the 1850s Ohio was the leading wine-producing state in the nation. But just in that decade the vineyards were attacked by black rot and mildew, which proved so virulent that small growers soon lost everything, and large growers were badly threatened. The reductions in manpower necessitated by the Civil War made things even worse, and by the end of the 1860s the wine industry in southern Ohio had ended. Today, all along the Ohio and especially toward Cincinnati, the industry is enjoying a remarkable revival. The northern river shore is again rich with grape growing.

Indiana's wine story began at about the same time, with John James Dufour's arrival in Kentucky in 1798. This immigrant from Vevey, Switzerland, established a variety of grapes around Lexington, Kentucky, but by 1810 had moved to southeast Indiana for its more favorable conditions and climate. He and his family established the town of Vevay, Indiana, growing vineyards there with grapes from the Cape of Good Hope and Madeira, via Philadelphia, where Dufour had purchased the vines. The "Cape" grapes may in fact have been a native-American *labrusca*, which looks similar. These varieties thrived in Vevay, where the family enjoyed bountiful harvests. Then, in the late 1820s, younger relatives lost interest and turned to more routine farm crops, and wine growing died out. The town still has an annual wine festival that celebrates the resurgence in local viticulture as well as the charm of its French-Swiss past. New Harmony, and its vicinity at the other end of the Ohio near the Wabash, also knew wine growing in the early nineteenth century under the care of a German religious sect led by George Rapp. But by the 1830s wine growing had, in this region too, disappeared for the most part. Viticulture has vigorously reappeared in recent years in the entire lower third of Indiana, due to initiatives by local growers and to the cooperation of the state legislature.

Ohio

HAFLE VINEYARDS / 2369 Upper Valley, Springfield, Ohio 45502
(513) 399-5742

OPEN MONDAY–THURSDAY NOON–8 P.M.; FRIDAY AND SATURDAY NOON–11 P.M.; CLOSED CHRISTMAS. WT, RS, P. FEE: 10¢ PER TASTE.

GR French hybrids; Baco; Villard; De Chaunac **R** Old Mill; Farm Restaurant **HM** Park Terrace; Holiday Inn **HS** U.S. Air Force Museum

MISC These vineyards were planted in 1970, and the winery has been constructed within a remodeled barn. Tasting booths are located in the old horse stalls. Hafle Vineyards features a summer market where the visitor can find homegrown fruits and vegetables, Amish cheeses, hams, honey, sorghum, etc. Grape stomps and wine festivals occur in early September, with German bands and other live music.

HAFLE VINEYARDS / 5010 South High Street, Columbus, Ohio 43215
(513) 399-2334

OPEN APRIL 1–DECEMBER 31 MONDAY–THURSDAY NOON–8 P.M.; FRIDAY AND SATURDAY NOON–11 P.M. WT, RS, P. FEE: 10¢ PER TASTE.

GR *Labrusca* **R** Old Mill; Farm Restaurant **HM** Park Terrace; Holiday Inn **HS** U.S. Air Force Museum.

MISC This thirty-acre vineyard was originally planted in 1921, and, like the Hafle Vineyards in Springfield, the winery here is in a remodeled barn. This vineyard also has a summer market offering local and fresh produce; and at several times during the summer the vineyard plays host to a day of wine tasting with Bratwurst accompaniment.

MEIER'S WINE CELLARS, INC. / 6955 Plainfield Pike, Cincinnati, Ohio 45236
Robert Raussen: (513) 891-2900

OPEN TUESDAY–SATURDAY 10 A.M.–1:30 P.M. WT, RS, D.

GR Catawba; Concord; Niagara; French hybrids; *vinifera*   **R** Meier's (at the winery); Grafton's   **HM** Ramada Inn; Hospitality Motor Inn.

MISC Meier's is the largest winery in Ohio, with an annual capacity of 2.5 million gallons, and some 350 acres under vines on Isle St. George in western Lake Erie. Champagne and Fruit, Table, Dessert, and Sparkling wines are produced here. The winery features its own restaurant, serving lunch and dinner, and during summer months there is outside garden dining.

MOYER VINEYARDS / Route 1, Manchester, Ohio 45144
Ken Moyer: (513) 549-2957

OPEN MONDAY–SATURDAY 11:30 A.M.–9:00 P.M. WT, RS, D. FEE MINIMAL.

GR French hybrids   **R** Moyer's (at the vineyards)   **HM** Fenner Motel; Brown Motel   **HS** Shawnee State Park.

MISC This young vineyard, established in 1973, produces six wines and a bottle-fermented Brut Champagne. These are offered to the visitor at the vineyards' own restaurant overlooking the Ohio River.

WARREN J. SUBLETTE WINERY / 2260 Central Parkway, Cincinnati, Ohio 45214
(513) 651-4570

OPEN MONDAY 10 A.M.–6 P.M.; TUESDAY–THURSDAY 10 A.M.–1 A.M.; FRIDAY AND SATURDAY 10 A.M.–2:30 A.M.; SUNDAY 6 P.M.–1 A.M.; CLOSED NEW YEAR'S DAY, THANKSGIVING, CHRISTMAS. WT, RS, D. FEE: $1.40–$1.75.

WP Chancellor; Marechal Foch; Baco Noir; Chelois Noir; Vee Port; Baco Rosé; Catawba Rosé; Seyval Blanc; Villard Blanc; Burdin Blanc; Vidal Blanc; Aurora   **R** Sublette Winery Restaurant; Scotti's; Yum Yum; Felafel House; Zino's Firehouse; Samuri   **HM** Carrousel Inn; Quality Inn; Marriott Inn; Treadway Inn   **HS** Cincinnati Art Museum; Ohio Historical Society; Dayton Street; Hauck House.

MISC Warren J. Sublette purchased this nineteenth-century brewery with its distinctive vaulted stone ceiling in 1975. The 125-year-old brewery has had a colorful history dating back to the days when Cincinnati was "The Champagne Capital of the New World." Since the mid-1800s the brewery has been used as a furniture store, a peanut processing factory, and an illegal distillery during Prohibition. Today, veteran wine maker Sublette has brought music, hospitality, and fine food to this venerable building to complement the wines he

THE OHIO RIVER VALLEY

proudly produces and serves. There is a restaurant serving a variety of lunches including ethnic specialties, sandwiches, cheeses, and fresh bread. Live folk music may be heard in the Stube Tuesday through Sunday evenings.

VALLEY VINEYARDS FARM, INC. / 2041 East US 22-3, Morrow, Ohio
45152   (513) 899-2485

OPEN MONDAY–THURSDAY 11 A.M.–8 P.M.; FRIDAY AND SATURDAY 11 A.M.–11 P.M.; CLOSED NEW YEAR'S DAY, THANKSGIVING, CHRISTMAS. WT, RS, P. FEE: 10¢ PER SAMPLE

GR Aurora; Seyval Blanc; Villard Blanc; Baco #1; De Chaunac; Foch; Chancellor; Blue Eye; Colobel; Chelois; Niagara; Concord; Catawba
R Golden Lamb Inn; King's Island Inn; International House   **HM** Golden Lamb Inn; King's Island Inn; Heritage Inn; Holiday Inn   **HS** Fort Ancient Park and Museum; Golden Lamb Inn.

MISC Since 1970 thirty acres of Valley Vineyards farmlands have been under vines. There is a patio with picnic tables and a party room with a fireplace. Pizza and cheeses are available.

VINTERRA FARM WINERY AND VINEYARD / 6505 Stoker Road,
Houston, Ohio 45333   John W. Monroe: (513) 492-2071

OPEN TUESDAY–THURSDAY 5 P.M.–9 P.M.; FRIDAY 4 P.M.–10 P.M.; SATURDAY 1 P.M.–10 P.M. WT, RS, P. FEE FOR TASTING BOARD (SEVEN WINES): $1.95.

GR Baco Noir; De Chaunac; Aurora; Vidal 256; Seyve-Villard 12-375; Chancellor   **R** Golden Lantern; Wooden Shoe Inn   **HM** Holiday Inn
HS Garth Museum; Ohio-Erie Canal Locks; Neil Armstrong Air and Space Museum; Old Northwest Museum; Wright-Patterson Air Force Museum; Johnston Farm Museum and Indian Agency.

MISC The vineyards at Vinterra Farm were planted in 1972; the winery was completed in 1977. The winery, in Bavarian style, offers an outdoor patio for wine tasting, with a scenic view of rolling countryside and farms. In 1974, five hundred apple trees were planted at Vinterra for future apple wine sale.

WYANDOTTE WINE CELLAR, INC. / 4640 Wyandotte Drive, Gahanna,
Ohio 43230   (614) 476-3624

OPEN MONDAY–SATURDAY 10 A.M.–5 P.M.; CLOSED LEGAL HOLIDAYS. CALL FOR APPOINTMENT. RS.

WP Grape; Rhubarb; Cherry. In production: Dandelion; Red Clover; Grapefruit; Blackberry; Blueberry; Tomato   **R** Wine Cellar; Garden Gallery   **HM** Sheraton Motor Inn North   **HS** Ohio Historical Society

Wyandotte Wine Cellar, Inc. (cont.)

MISC Jointly owned and operated by Floyd Jones and George Kraus, the Wyandotte Wine Cellar is an offshoot of the Amana, Iowa, winery started by George Kraus thirty years ago.

Indiana

EASLEY WINERY / 205 North College Avenue, Indianapolis, Indiana 46202   **(317) 636-4516**

OPEN MONDAY–SATURDAY NOON–6 P.M.; CLOSED LEGAL HOLIDAYS. WT, RS.

WP Varietal; generic; specialties   **R** La Scala; St. Elmo; La Tour; King Cole; Red Lobster; Laughner's Cafeteria   **HM** Hyatt Regency; Hilton; Holiday Inn; Howard Johnson's; TraveLodge; Atkinson; Stouffer's   **HS** Indianapolis Museum of Art; Children's Museum; Memorial Plaza; Monument Circle; President Harrison's Home.

MISC The owners of the Easley Winery, located in downtown Indianapolis, are professional people, and describe their winery as "a mom and pop operation all the way." The winery's private vineyard is located some forty miles downriver from Louisville, Kentucky, on the Ohio River.

GOLDEN RAIN TREE WINERY / R.R. 2, Wadesville, Indiana 47638   **(812) 963-6441**

OPEN MONDAY–SATURDAY 8 A.M.–10 P.M. WT (FEES VARY); RS, P, D.

WP Table wines   **R** Dutch Corner   **HM** Ramada Inn   **HS** Town of New Harmony.

MISC Founded in 1975, this young winery also has fresh grape and apple juice, in season, available for home wine makers. Personalized wine labels are also available.

THE OHIO RIVER VALLEY

OLIVER WINERY / 8024 North Hiway 37, Bloomington, Indiana 47401
(812) 876-5800

OPEN MONDAY–SATURDAY 11 A.M.–6 P.M.; CLOSED NEW YEAR'S DAY, THANKSGIVING, CHRISTMAS. WT, RS, P; D LIGHT CHEESE LUNCHEONS.

WP Dry, Medium, Semi-sweet, and Honey wines   **R** Fireside Inn; Gold Rush; Inn of the Four Winds   **HM** Fireside Inn; Holiday Inn; Inn of the Four Winds   **HS** Art and History museums at the University of Indiana.

MISC Bill Oliver is a professor of law at Indiana University, as well as the owner of this winery, founded in 1972. The first plantings at his vineyards were in 1966.

POSSUM TROT VINEYARD / Unionville, Indiana 47468
Ben and Lee Sparks: (812) 988-2694

OPEN APRIL–DECEMBER MONDAY–SATURDAY 10 A.M.–SUNSET. TOURS BY APPOINTMENT. WT, RS, P; D CHEESE AND BREAD. FEE: $1.00.

GR French hybrids   **HM** Ramada Inn; Brown County Inn   **HS** Brown County Art Gallery and State Park; Old Log Jail; County Museum; T.C. Steele State Memorial.

MISC Owner Ben Sparks was a pioneering wine grower in Indiana when he planted his first vines in 1967. He is the organizer of the annual Indian Grape Wine Symposium, held in Nashville, Indiana, during the month of March. Mr. Sparks writes that the beauty of the scenery "makes the trip to Possum Trot Vineyards an unforgettable experience—once you find it."

SWISS VALLEY VINEYARD / 101 Ferry Street, Vevay, Indiana 47043
Alvino F. Meyer: (513) 521-5096

OPEN BY APPOINTMENT ONLY. WT, RS, P. FEE: 10¢.

GR French hybrids   **R** Swiss Inn   **HM** Swiss Inn.

Kentucky

COLCORD WINERY / Third and Pleasant Streets, Paris, Kentucky 40361
(606) 987-7440

OPEN APRIL 16–DECEMBER 31 TUESDAY–SATURDAY 11 A.M.–6 P.M.; OR BY APPOINTMENT. WT, RS.

GR French-American hybrids; *vinifera*   **HS** Cane Ridge Meeting House; Duncan Tavern.

MISC It was not until August 1976 that Kentucky legislation permitted small wineries to operate legally in that state. The Colcord Winery, owned by F. Carlton Colcord, Jr., was the first Kentucky winery to be licensed.

The Great Lakes Region (Minnesota, Wisconsin, Michigan, Illinois, Indiana, Ohio, Pennsylvania, New York, and Ontario)

Wineries located near the Great Lakes share a condition uniquely suited for the growing of grapes—the climate along these shores is tempered by what is known as the "Lake Effect." The great body of water nearby retains heat from the summer sun, so that autumns are mild, and the growing season prolonged, as the lake feeds its accumulated warmth into the earth. Near these lakeshores the temperature is 5 to 10 degrees warmer in winter and cooler in summer than temperatures farther inland. Vintners in Minnesota, Wisconsin, Illinois, Indiana, Ohio, Pennsylvania, and New York Great Lakes regions take advantage of this very beneficial grape weather, and of hills and slopes fed by lake breezes, to grow a wide variety of hybrid and native grapes. A trip along the lakeshores in these states will take you through some very wine-conscious communities and afford you a good selection of tastes.

In many of these states the Lakes-country wine tradition is more than a hundred years old, with successful wine production dating back to the 1850s. The history of growing in areas like the Lake Erie Islands near Sandusky, Ohio, will provide an interesting background to the vintages produced there today. These islands, with the longest growing season of any eastern wine region, are a picturesque place to visit, and have grown distinguished vintages since the 1840s. At Nauvoo, Illinois (on Highway 96), the Nauvoo

State Historic Site and Park contains two acres of the original vineyards planted in 1851 by two German immigrants, Alois Reinberger and the Reverend Alleman. Citizens and visitors are invited to participate in the grape harvest on the site in late August, and to visit the adjacent wine museum, open all year round, which displays over thirteen hundred wine artifacts.

Unlike more eastern areas, Michigan, with little history as a wine-producing region, is experiencing what the Michigan Wine Institute calls a "Wine Rush." Since the early 1950s a great proliferation in Michigan wine growing has taken place. There is much experimentation with grape varieties by Michigan growers; they have found, for instance, that hybrid varieties are doing even better by the Lakes than the native grapes which have always grown there. Michigan vintners are also experimenting with European varieties, of which the white grape stocks are the most successful in the cool climate. Growers like the climate, the soil, and the slope and roll of the lakeside terrain, and it is felt that Michigan's recent beginnings in wine production will yield some very interesting results indeed. The telephone number of the Michigan Wine Institute in Detroit is (313) 886-4629, and their address is P.O. Box 05204, Detroit, Michigan 48205.

The combination of recently established vineyards and older wine districts makes quite an interesting series of tours possible along the Great Lakes shores region. The "Lake Effect" assures you of interesting wine-tasting opportunities all along the way.

Minnesota

ALEXIS BAILLY VINEYARD, INC. / 18200 Kirby Avenue, Hastings, Minnesota 55033 (612) 437-1413

OPEN MAY–NOVEMBER TUESDAY–SATURDAY 9 A.M.–5 P.M. WT, RS, P.

GR Foch; Millot; Seyval Blanc; Minnesota hybrids   **R** Various restaurants in the Hiawatha Valley and Twin Cities area.

MISC The vineyards at Alexis Bailly were planted in 1973, and the "craft style" winery was built in 1976. The wines here are made exclusively from Minnesota fruits.

LAKE SYLVIA VINEYARD / Route 1, South Haven, Minnesota 55382
NOT OPEN TO THE PUBLIC. RS BY APPOINTMENT.

GR French hybrids; *vinifera;* locally developed varieties.

MISC Owner David W. Macgregor notes that his vines must be taken off the trellis and covered to protect them from the winter temperatures, which run as low as forty degrees below zero. The first crush was in 1976.

Wisconsin

STONE MILL WINERY, INC. / N. 70, W. 6340, Bridge Road, Cedarburg, Wisconsin 53012   **(414) 377-8020**
OPEN MONDAY–SATURDAY 10 A.M.–5 P.M.; SUNDAY NOON–5 P.M.; CLOSED NEW YEAR'S DAY, EASTER, CHRISTMAS. WT, RS, P, D. FEE FOR TOUR AND TASTING: $1.00 PER ADULT.

WP Natural Cherry; Special Natural Honey Cherry; Special Natural Colonial Spice   **R** Cream and Crepes Cafe; Adam's Rib   **HM** Wulff's Island; Harbor Shore   **HS** Hamilton Historic District; historic downtown Cedarburg; Pioneer Village.

MISC In 1972 vintner James D. Pape began restoration of an 1864 woolen mill which overlooks Cedar Creek. Here he located his winery, which now produces up to fifteen thousand gallons of wine annually. The winery is in the vicinity of several specialty shops and restaurants.

WOLLERSHEIM / Highway 188, Prairie du Sac, Wisconsin 53578
Robert or JoAnn Wollersheim: (608) 643-6515

OPEN DAILY 10 A.M.–5 P.M. WT, RS, P. FEE FOR HISTORICAL TOURS: $1.00 AGE TWELVE AND OVER.

GR White Riesling; Pinot Chardonnay; French hybrids   **R** Firehouse Restaurant   **HM** Ganser's Motel; Black Hawk Ridge; Skyview Motel   **HS** St. Norbert Church.

MISC On his way westward, Count Agoston Haraszthy, "the father of California wine making" founded and lived in Prairie du Sac/Sauk City, Wisconsin, long enough to plant the original vineyards here. When the count moved on, Peter Kehl built the current winery building and family house in 1858. It was in 1972 that the Wollersheim Family reopened the winery and replanted the Haraszthy vineyards. Today wine maker Robert P. Wollersheim is extremely active in revivifying Wisconsin's interest in wine. Historical tours of the vineyards and winery are available, as are occasional seminars in home wine making and grape growing. The winery offers a gift shop with a large selection of wines from around the world as well as a variety of glassware and books on wine. There is a wide selection of cheeses, wine and cheese gift boxes, and books and supplies for the home wine maker.

Michigan

BOSKYDEL VINEYARD / Route 1, Box 522, Lake Leelanau, Michigan 49653
Bernard C. Rink: (616) 256-7272

OPEN DAILY 1 P.M.–6 P.M.; CLOSED NEW YEAR'S DAY, EASTER, THANKSGIVING, CHRISTMAS. FREE TOURS AND TASTING ROOM. CALL FOR APPOINTMENT FOR SPECIAL GROUP OR BUS TOURS AT ODD HOURS.

GR French-American hybrids   **R** Leland Lodge; Bluebird   **HM** Park Place Motor Inn; Leland Lodge   **HS** Sleeping Bear Dunes National Park.

MISC Wine maker Bernard Rink began testing hybrid grapes in 1965, and now he, his wife, and sons work the Boskydel Vineyards. Geographically Mr. Rink likens the site of his vineyards to such famous wine-growing areas as the Barossa Valley of Australia, the Rhine and Mosel regions of Germany, and the Finger Lakes district of New York State.

THE GREAT LAKES REGION   111

BRONTE WINERY AND VINEYARDS / Keeler, Michigan 49057
(616) 621-3419

OPEN MONDAY–SATURDAY 10 A.M.–4 P.M.; SUNDAY NOON–4 P.M.; CLOSED NEW YEAR'S DAY, EASTER, THANKSGIVING, CHRISTMAS. WT, RS.

GR Baco Noir; De Chaunac; Marechal Foch; Delaware; Catawba; Aurora Blanc; Vidal Blanc; Vee Port   **HM** Holiday Inn.

MISC The Bronte Vineyards are tied to the past in ways that are varied, and sometimes amusing. It was, for example, at a small farmhouse still standing at the vineyards that the Keeler branch of the Women's Christian Temperance Union was formed in 1879. Several of the largest wood wine-storage tanks used today were originally used by Al Capone in one of his breweries. And it was the Bronte Winery which first bottled Cold Duck. The Bronte Champagnes and wines have now received international acclaim, and today this is the largest manufacturer of premium Champagne and wine in Michigan. Free tours are offered, children are welcome, and the hospitality room at the winery features a magnificent view and free wine tasting.

CHÂTEAU GRAND TRAVERS, LTD. / 12239 Center Road, Traverse City, Michigan 49684   (616) 333-7355

OPEN MONDAY–FRIDAY 10 A.M.–4 P.M. WT, RS, P.

GR Riesling; Chardonnay; Cherry; Apple   **HS** Old Mission.

MISC Wine maker Karl Werner came to Michigan with impressive credentials. His family has been producing wine in Germany since 1411, and at one time he was in charge of the German Government Wine Cellars. He has a master's degree in enology and viticulture, and is an associate professor of enology at the University of California. He is a consultant to wineries in Europe and in California, and has worked for Mondavi and Gallo. He is, to a large extent, responsible for the White wine industry in South Africa. While Traverse City is sometimes known as "the Cherry Capital of the World," Werner believes he can establish a truly great White wine district here as well. Château Grand Travers planted its first vines in 1976.

FENN VALLEY VINEYARDS AND WINE CELLAR / R.R. #4, 6130 122nd Ave. at 62nd St., Fennville, Michigan 49408
William W. Welsch: (616) 561-2396

OPEN MONDAY–SATURDAY 10 A.M.–5 P.M.; SUNDAY 1 P.M.–5 P.M.; CLOSED NEW YEAR'S DAY, THANKSGIVING, CHRISTMAS. CALL AHEAD. WT, RS, P.

GR White Riesling; Gewürztraminer; European hybrids; also cherries and peaches   **R** Tara; Coral Gables; Butler; Skyline; Jocko's; West Shore Golf Club; Clearbrook Golf Club; Hubbard House; Point West   **HM** Amity Motel;

Fenn Valley Vineyards and Wine Cellar (cont.)

Shangrila Motel; Ship 'n' Shore Motel and Boatel; Timberlane Motel; Saugatuck Lake Shore; Plaza Motel; Holiday Inn; Wooden Shoe Motel **HS** Grand Haven.

MISC After a three-year search the Welsch Family established their vineyard in Fennville in 1973. Only those grapes best suited to the severe climate of the Great Lakes Region were planted. Being located in a prime fruit area, Fenn Valley is also suitable for producing outstanding *true Fruit* wines. The tasting room features an elevated observation deck overlooking the cellars, and audiovisual displays about the use of wine. Cold plate lunches available at the winery may be enjoyed in the adjoining picnic area.

FINK WINERY / 208 Main Street, Dundee, Michigan 48131
                                                                                    (313) 529-3296

OPEN MONDAY–SATURDAY 10 A.M.–4 P.M.; CLOSED MAJOR HOLIDAYS. WT, RS.

WP Red and White Table wines; Rosé, Rubyred-like-Lambrusco; Mead; Concord.

MISC This small winery was founded in 1976.

FRONTENAC VINEYARDS, INC. / P.O. Box 215, 3418 West Michigan Avenue, Paw Paw, Michigan 49079   Carl Corsi: (616) 657-5531

OPEN MONDAY–SATURDAY 9 A.M.–5 P.M.; SUNDAY NOON–5 P.M.; CLOSED NEW YEAR'S DAY, EASTER, CHRISTMAS. WT, RS.

GR Concord; Niagara; Fredonia; hybrids   **R** Gateway Restaurant; Gene Mitchell's Restaurant; Sugar Bear Restaurant; Di Juanco's   **HM** Holiday Inn; Ramada Inn; Green Acres Motel.

MISC Frontenac was established at the time of Repeal in 1933. The Frontenac wines have been awarded prestigious international and American gold medals. The visitor will find campgrounds, antique shops, and a flea market located nearby.

LAKESIDE VINEYARD, INC. / 13581 Red Arrow Highway, Harbert, Michigan 49115   Helen Lundquist: (616) 469-0700

OPEN MONDAY–SATURDAY 9 A.M.–5 P.M.; SUNDAY NOON–5 P.M.; CLOSED NEW YEAR'S DAY, EASTER, THANKSGIVING, CHRISTMAS. TOURS APRIL 16–NOVEMBER, MONDAY–SATURDAY 11 A.M.–4 P.M.; SUNDAY NOON–4:30 P.M. WT, RS, P.

THE GREAT LAKES REGION 113

GR Niagara; Concord   **R** Tosi's; Win Schuler's; Woodshed; Pebblewood Country Club; Little Bohemia; Scotty's   **HM** Lazy V. Motel; Holiday Inn; Ramada Inn; Howard Johnson's   **HS** Warren Dunes State Park; Cook Nuclear Center.

MISC Special attractions at the Lakeside Vineyard include: a May Festival (last weekend in May); an October *Weinfest* (first weekend in October); and an Art Show (third weekend in August). Tours are conducted by guides dressed in traditional costumes of the Revolutionary period, in keeping with the Molly Pitcher line of wines produced here. During the summer months a newly built wine garden is open to the public.

LEELANAU WINECELLARS, LTD. / 726 North Memorial Highway, U.S. 31, Traverse City, Michigan 49684   Marian A. Haw: (616) 946-1653
OPEN DAILY 10 A.M.–6 P.M. WT, RS; P NEARBY.

WP Chardonnay; Merlot; Cabernet Sauvignon; Seyval Blanc; Aurora; De Chaunac; Chelois; Baco Noir; Leelanau Red; White; Rosé; Apple; Pear; Cherry; Cherry Rosé; Strawberry; Nectarine.

MISC This Traverse City winery is owned and operated in conjunction with the larger Leelanau Winery and Vineyards, located thirty miles away in Omena, Michigan. Although the larger winery is not open to the public, all wines made there are available at the Traverse City tasting room. A special new feature at the attractive tasting room is the 100 percent Nectarine wine. Children are welcome, and there is a playground next to the winery.

L. MAWBY VINEYARDS AND WINERY / 4519 Elm Valley Road, Suttons Bay, Michigan 49682   Lawrence Mawby: (616) 271-3522
OPEN SATURDAY AND SUNDAY NOON–6 P.M. BY APPOINTMENT; CLOSED LEGAL HOLIDAYS. WT, RS, P.

GR French-American hybrids   **R** Rowe Inn   **HM** Holiday Inn; Park Place.

MISC These vineyards, planted in 1974, had their first commercial crush in 1978.

ST. JULIAN WINE COMPANY, INC. / 716 South Kalamazoo Street, Box 112, Paw Paw, Michigan 49079   (616) 657-5568
OPEN MONDAY–SATURDAY 9 A.M.–5 P.M.; SUNDAY NOON–5 P.M.; CLOSED NEW YEAR'S DAY, EASTER, THANKSGIVING, CHRISTMAS. WT, RS.

St. Julian Wine Company, Inc. (cont.)

WP Table, Dessert, and Sparkling wines; Champagnes; Flor Sherry; grape juice and cooking wines **R** La Cantina; Gateway Inn; Gaspare's Place **HM** Holiday Inn; Kalamazoo Hilton; Sheraton; Ramada Inn **HS** Western Michigan University; Nazareth College; Kellogg's Battle Creek.

MISC Founded in Canada in 1921, the St. Julian Wine Company, the oldest in Michigan, moved to Paw Paw in 1936. Owner Mariano Meconi developed the wine company, and now St. Julian wines are sold throughout the Midwest. Extensive work has been done in the last decade to expand production and upgrade quality. The wine company was awarded Michigan's governor's "Ambassador of Tourism Award" in 1976 for being the biggest contributor to Michigan tourism that year. The hospitality room features a cheese shop and deli, and the winery, with advance notice, caters to group tours and busloads.

TABOR HILL VINEYARD AND WINECELLAR, INC. / Route 1, Box 746, Mt. Tabor Road, Buchanan, Michigan 49107
Leonard R. Olson: (616) 422-1161

OPEN DAILY NOON–5:30 P.M. WT, RS, P; D (SPECIAL GROUP RATES). FEE FOR EVENING TOURS: $3.00 PER PERSON; SENSORY WINE EVALUATION: $5.00.

WP Chardonnay; Johannisberg Riesling; Cuvee Rosé; Cuvee Rouge; Cuvee Blanc; Baco Noir; Vidal Blanc 75; Vidal Blanc 76; Vidal Blanc Sec **R** Win Schuler's **HM** Holiday Inn; Ramada Inn; Howard Johnson's.

MISC Formerly affiliated with a major manufacturer of steel, wine maker Leonard Olson founded the Tabor Hill Vineyard in 1968. In 1977 Tabor Hill produced over 25,000 cases of their various wines, some of which were served by President Ford at the White House. According to Olson, "Although the goal of making some of America's finest wine seems like a romantic dream, frankly, it has taken one hell of a lot of work." Olson notes that wine has always been associated with art, literature, and music. In keeping with this philosophy he is proud to display the work of local artists at the winery. A different artist is featured every six weeks.

VENDRAMINO VINEYARDS COMPANY / Route 2, Paw Paw, Michigan 49079
Harriet Doty: (616) 657-5890

OPEN DAILY APRIL–NOVEMBER 11 A.M.–5 P.M. WT, RS, P.

WP Red, White, Rosé, and specialty wines **R** Gateway Inn; Di Juanco's **HM** Green Acres-Motel.

THE GREAT LAKES REGION

MISC Owner and wine maker John Coleman had been interested in home wine making since 1971. His own winery was bonded in 1976, and his Red Table wine was awarded a gold medal at the Michigan State Fair in 1977. The Vendramino Vineyards have playground equipment for children, and offer cider and doughnuts during the fall months.

WARNER VINEYARDS / 706 South Kalamazoo Street, Paw Paw, Michigan 49079   Richard Palasinski: (616)-657-3165

OPEN MONDAY–SATURDAY 9 A.M.–5 P.M.; SUNDAY NOON–5 P.M.; CLOSED NEW YEAR'S DAY, EASTER, THANKSGIVING, CHRISTMAS. TOURS MONDAY–SATURDAY 10 A.M.–4 P.M.; SUNDAY 1 P.M.–4 P.M. WT, RS; P NEARBY.

GR Native American; French hybrids; some *vinifera*   **R** Di Juanco's; Sveden House   **HM** Green Acres Motel; Holiday Inn   **HS** Van Buren County Museum; Kalamazoo Museum; Museum of the City of Marshall.

MISC Warner Vineyards, founded in 1938, is owned and operated by the Warner Family. The winery has a hospitality center which is called the Wine Haus. It is located in the former village water works building which was built in 1898. A tour consists of a twelve-minute slide presentation and a thirty-minute walking tour of the winery. The owners comment: "A tour through Warner Vineyards is a learning experience the whole family will enjoy. We hope you will take the time to visit us."

Illinois

GEM CITY VINELAND COMPANY, INC. / East Parley Street, Nauvoo, Illinois 62354   (217) 453-2218

OPEN DAILY 8 A.M.–5 P.M. WT, RS.

GR Concord; Niagara   **R** Icarian Lounge   **HM** Hotel Nauvoo   **HS** Mormon Church Restoration.

MISC This winery has been in the Baxter Family since 1857. The winery participates in the Nauvoo "Wedding of Wine and Cheese" every Labor Day weekend.

MOGEN DAVID WINE CORP. / 3737 South Sacramento Avenue, Chicago, Illinois 60632
(Executive Offices: 444 North Michigan Avenue, Chicago, Illinois 60611)
Earl Engus: (312) 254-6300

OPEN MONDAY–FRIDAY BY APPOINTMENT ONLY; CLOSED LEGAL HOLIDAYS, JEWISH HOLIDAYS, AND DURING THE MONTH OF JULY. WT.

WP Eastern varieties; Red; White; Rosé.

MISC The Mogen David story is one of continued growth, expansion, and success. This was one of the few wineries allowed to operate during Prohibition. It began in a $45-a-month loft in downtown Chicago, bottling wine sold primarily to churches and synagogues for sacramental purposes. Today Mogen David wines can be found throughout the world. The views of John Koche, wine maker for Mogen David, are expressed in a recent press release: "Wine making has come a long way from the little-old-wine-maker days of stomping by foot and stirring by hand. Because it's such a controlled scientific process, it's easy to see why wineries in the heart of Chicago can produce wines as sweet and luscious as if they were right next door to the vineyards." Mogen David wines are acceptable for Kosher use.

NAUVOO STATE HISTORIC SITE AND PARK VINEYARD / P.O. Box 337, Nauvoo, Illinois 62354   Bob Cashman: (217) 453-2512

OPEN DAILY APRIL–OCTOBER 9 A.M.–5 P.M.

GR Concord   **R** Hotel Nauvoo Dining Room; Kraus Cafe and Restaurant; Icarian Lounge; Press House Restaurant; Sloan's Restaurant   **HM** Hotel Nauvoo; Pioneer Motel; Village Inn Motel; Best Western Motel; Holiday Inn   **HS** The city of Nauvoo features homes restored in the style of the Mormon period (1839–1846) and the Icarian Period (1849–1856). The local chamber of commerce offers tours of some thirty restored homes and shops.

MISC Alois Reinberger and the Reverend Alleman are credited with planting the first grapevines in Nauvoo in 1851. Of this initial planting, two acres survive today at the Nauvoo Historic Site and Park. These are tended by Park personnel and harvested by the general public upon ripening in late August. Although these grapes are not made into wine, there is a wine-making museum adjacent to the vineyards, exhibiting over thirteen hundred artifacts, and a wine cellar of the 1850s.

THE GREAT LAKES REGION 117

THOMPSON WINERY / Illinois Route 50 and Pauling Road, Monee, Illinois 60449

OPEN SUNDAY 1 P.M.–4 P.M. WT, RS, P, D. FEE: $1.00 PER ADULT.

WP Pere Marquette Brut, Pink, and Natural.

MISC The winery features a wine-making supply store.

Indiana

BANHOLZER WINECELLARS, LTD. / 5627 East County Route 1000 North, New Carlisle, Indiana
Carl Banholzer or Judith Braun: (219) 778-2448

OPEN MONDAY–SUNDAY 11 A.M.–5 P.M.;CLOSED NEW YEAR'S DAY, CHRISTMAS. WT, RS, P, D.

GR Cabernet Sauvignon; Pinot Noir; Chardonnay; Seyval Blanc; Gamay Beaujolais; Vignoles; Baco Noir; Foch; Vidal Blanc; Villard Blanc; White Riesling   **R** Hesston Bar; Tinkers Dam; Tangerine; Timbers Restaurant   **HM** Holiday Inn.

MISC Although open to the public, Banholzer Winecellers has a private membership of some five thousand and publishes its own newsletter advising members of special events at the vineyards. The first plantings on these seventy-two acres were in 1971, and Banholzer now features a "Wine Art Gallery," a gourmet food shop, and a delicatessen with imported cheeses and freshly baked bread. There are five acres of picnic area, and occasional swimming, fishing, and flatbed tours of the vineyards' private lake.

RAUNER AND SONS WINECELLARS AND VINEYARD / 314 Dixieway North, South Bend, Indiana 46637   James Rauner: (219) 277-4078

OPEN MONDAY–SATURDAY 11 A.M.–7 P.M. WT, RS, P.

GR French hybrids; *vinifera*   **R** Howard Johnson's; Steak and Ale; Boars Head   **HM** Ramada Inn   **HS** University of Notre Dame.

MISC This newly established vineyard and wine cellar boasts a gift shop which sells wine- and beer-making supplies, books, glasses, and other miscellaneous gifts. The store is open Monday through Saturday, 11 A.M.–7 P.M.

118   THE TRAVELER'S GUIDE TO THE VINEYARDS OF NORTH AMERICA

VILLA MEDEO VINEYARDS / Aulenbach Avenue, Route 2, Madison, Indiana 47250 (812) 265-2144

OPEN MONDAY–SATURDAY 11 A.M.–5 P.M. WT, RS, P. FEE: VARIABLE.

GR French-American crosses   **R** Key West Shrimp House   **HM** Hillside Inn; President Madison Hotel; Victoria Inn   **HS** Madison features six restored antebellum museum homes.

MISC The winery was bonded in 1974.

Ohio

AU PROVENCE RESTAURANT AND CEDAR HILL WINE COMPANY / 2195 Lee Road, Cleveland Heights, Ohio 44118
Richard Taylor or Dr. Thomas Wykoff: (216) 321-9511

RESTAURANT: OPEN MONDAY–FRIDAY 6 P.M.–10 P.M.—NO RESERVATIONS; SATURDAY 6:30 P.M. AND 9:15 P.M. SEATINGS ONLY—RESERVATIONS NECESSARY.

GR *Vinifera;* French-American hybrid   **WP** Table wines   **R** Au Provence; Earth by April; Greenhouse   **HM** Park Plaza; Holiday Inn; Marriott Inn Airport; Hollinden House   **HS** Cleveland Museum of Art; Crawford Auto Museum; Cleveland Orchestra.

MISC *Holiday* magazine has given an award to the Au Provence Restaurant for its excellence. It is a unique restaurant with its own basement winery to produce wines solely for Au Provence. A retail store is now open next door to the restaurant. Founded in 1974, Cedar Hill Wine Company produces 5,200–6,000 gallons of wine each year.

CHALET DEBONNÉ VINEYARDS, INC. / 7743 Doty Road, Madison, Ohio 44057 Anthony P. Debevc: (216) 466-3485

OPEN TUESDAY, THURSDAY, SATURDAY 1 P.M.–8 P.M.; WEDNESDAY, FRIDAY 1 P.M.–MIDNIGHT; CLOSED NEW YEAR'S DAY, THANKSGIVING, CHRISTMAS. WT, RS; D (LIMITED).

GR *Labrusca*; French hybrids; *vinifera*   **R** Old Tavern; Geneva Inn   **HM** Howard Johnson's; Quail Hollow; Holiday Inn.

THE GREAT LAKES REGION 119

MISC The Debevc Family has been making and sharing wines for three generations, but it was not until 1970 that they turned to commercial wine making. Today two generations tend the forty vineyard acres and operate the underground winery. Above the winery is a tasting room done in the style of a Swiss chalet. A visitor or family can sample wines which may be complemented by cheese, sausage, and homemade bread. In the evening the Debevc family appreciates a touch of formality, and asks that visitors not wear blue jeans after 8 P.M.

DOVER VINEYARDS / 24945 Detroit Road, Westlake, Ohio 44145
(216) 871-0700

OPEN MONDAY–SATURDAY 9 A.M.–5:30 P.M.; CLOSED MAJOR HOLIDAYS. WT, RS, D.

WP Concord; Catawba; Niagara; and six Champagnes **R** Dover Chalet.

MISC This winery was originally constructed in the 1930s by a grape-growers' cooperative, as a means of selling extra grapes. The current owner, Zoltan Wolovits, bought the winery from the cooperative and built the Dover Chalet Restaurant. The vineyard features a home beer- and wine-making-supply store as well.

E AND K WINE COMPANY / 220 East Water Street, Sandusky, Ohio 44870
Clifford J. Gregory: (419) 627-9622

OPEN MONDAY–SATURDAY 10 A.M.–6 P.M.; CLOSED JANUARY–APRIL AND MAJOR HOLIDAYS. WT, RS, D. FEE: $1.00 PER PERSON, REGULAR; $2.00, DELUXE.

WP Native American **HS** The E and K Wine Company Building is listed in the National Register of Historic Sites.

MISC This winery, established in 1863, was at one time the largest winery east of the Mississippi, with a capacity of 865,000 gallons. The winery was reopened in 1971 and now has an annual production of 5,000 gallons.

GRAND RIVER VINEYARD / 5750 Madison Road, Madison, Ohio 44057
(216) 298-9838

OPEN APRIL 1–DECEMBER 31 MONDAY–SATURDAY 1 P.M.–SUNSET. WT, RS, P. FEE: VARIABLE.

GR Chardonnay; Gamay Beaujolais; Pinot Noir; Sauvignon Blanc; Merlot; Vignoles; Dutchess; Chancellor; Millot; Landot 4511; Chambourcin
R Quail Hollow Inn; Swallows; Unionville Tavern **HM** Quail Hollow Inn; Holiday Inn **HS** Mormon Temple; Garfield Home; Mentor Headlands State Park; Burton Historic Village; Shandy Hall Historic House.

Grand River Vineyard (cont.)

MISC The first vines at Grand River were planted in 1972. Until 1976, when the Grand River cellar was built, the grape harvests were sold to other wine makers. Since some of these other wines were awarded prizes, owner and wine maker Bill Worthy feels confident that his own wines will succeed as well.

HEINEMAN WINERY / Put-in-Bay, Ohio 43456
Louis V. Heineman: (419) 285-2811

OPEN APRIL 1–NOVEMBER 15 BY APPOINTMENT ONLY. WT, RS, P. FEE: 40¢ PER GLASS.

GR Catawba; Concord; Niagara; Ives; Delaware   R Coopers; Boat House   HS Perry Victor and International Peace Memorial

MISC This winery, founded in 1888 by Gustav Heineman, is now owned and operated by the founder's grandson. Louis Heineman offers tours of his winery and of Crystal Cave (on the vineyard grounds).

KLINGSHIRN WINERY / 33050 Webber Road, Avon Lake, Ohio 44012
Allan A. Klingshirn: (216) 933-6666

OPEN MONDAY–FRIDAY 1 P.M.–7 P.M.; SATURDAY 9 A.M.–7 P.M.; CLOSED NEW YEAR'S DAY, THANKSGIVING, CHRISTMAS. TOURS BY APPOINTMENT ONLY. WT, RS.

GR Concord; Niagara; Delaware   R Aquamarine; The Shaft at the Landings; Old English Parlor; Chef Henri (restaurant with local playhouse)   HM Saddle Inn; Aquamarine Motel.

MISC Owner and wine maker Allan A. Klingshirn purchased this winery from his father, Albert Klingshirn, who founded the winery in 1935, soon after Repeal. The original winery was located in the old cellar of his home and he produced one thousand gallons of Dry Concord each year. The Klingshirn Winery is still a one-man operation, with an annual production of approximately ten thousand gallons.

MANTEY VINEYARDS, INC. / 917 Bardshar Road, Sandusky, Ohio 44870
Paul Mantey: (419) 625-5474

OPEN MONDAY–FRIDAY BY APPOINTMENT ONLY, 9 A.M.–5 P.M.; SATURDAY 9 A.M.–NOON. WT, RS.

GR Catawba; Baco Noir; Seibel Chelois; Concord; Pinot Chardonnay
WP Blue Face Concord; Rare Ruby Port.

THE GREAT LAKES REGION 121

MISC Paul Mantey is proud of the fact that this family business has been at the same location since 1880. His Creme Sherry was the 1978 gold medal winner, Eastern United States.

MARKKO VINEYARD / R.D. #2 South Ridge Road, Conneaut, Ohio 44030
Arnulf Esterer: (216) 593-3197

OPEN MONDAY–SATURDAY 11 A.M.–6 P.M. WT, RS, P. FEE: $1.25 PER PERSON.

GR Johannisberg Riesling; Chardonnay; Cabernet Sauvignon   **R** Hil-Mak Seafood Restaurant; Olde Kings Tavern; Pape's   **HM** Ramada Inn   **HS** Conneaut Railroad Museum; Fort Independence; antique exhibits from the steam era.

MISC Founded in 1968 by Arnulf Esterer and Thomas Hubbard, the Markko Vineyard features vines grafted from those of Dr. Konstantin Frank (see listing, Finger Lake, New York, K. Frank). Annual production is approximately two thousand cases of which 80 percent is Riesling and Chardonnay, and the remainder Cabernet Sauvignon. New releases are announced by mail to members of the Case Club. Visitors may ask to be put on the mailing list.

STEUK WINE COMPANY / 1001 Fremont Avenue, Sandusky, Ohio 44870
(419) 625-0803

OPEN DAILY; CLOSED NEW YEAR'S DAY, CHRISTMAS. RS, P.

GR Black Pearl; Montefiore; Elvira; Mountain Catawba; Concord; Clinton; Foch   **R** Brown Derby; Le Gourmet; Four Monks; Mon Ami   **HM** Ramada Inn.

MISC The Steuk Family has been active in viticulture since the mid-1800s. When Prohibition forced the closing of their winery they had some sixty acres in vines. The present activity at the winery is credited primarily to the late William K. Steuk, who restored the winery after Repeal, placing great emphasis on the utilization of native-American grapes.

WILLOUGHBY WINERY / 30829 Euclid Avenue, Wickliffe, Ohio 44092
(216) 943-5405

OPEN TUESDAY–SATURDAY. WT, RS, D.

GR *Labrusca*; French hybrids.

MISC Willoughby Winery houses La Vigneta, a family restaurant featuring seafood and Italian food.

Pennsylvania

**LAPIC WINERY / 682 Tulip Drive, New Brighton, Pennsylvania
(412) 846-2031**

OPEN TUESDAY–SATURDAY 11 A.M.–9 P.M.; CLOSED NEW YEAR'S DAY, ELECTION DAY, CHRISTMAS. WT, RS.

GR French hybrids; *labrusca* **R** Hilltop Restaurant **HS** Merrick Free Art Gallery; Old Economy; Little Beaver Museum; Brodhead Cultural Center; Mill Creek Valley Historical Museum; Richmond Little Red Schoolhouse.

MISC Located considerably south of the North East lake vineyards of Pennsylvania, Lapic claims to be the first and only winery located in southwest Pennsylvania. This is a family winery, offering hospitable tours and tastings to the visitor.

MAZZA VINEYARDS / 11815 East Lake Road, North East, Pennsylvania 16428 Robert or Frank Mazza: (814) 725-8695

OPEN MONDAY–SATURDAY 9 A.M.–5:30 P.M.; CLOSED MAJOR HOLIDAYS AND ELECTION DAY. WT, RS, P. FEE: VARIABLE.

GR Riesling; Chardonnay; Cabernet; French and American hybrids **R** Delhurst Country Inn; Mums Motel **HM** Paschke's Mums Motel; Holiday Inn; Ramada Inn.

MISC The wine maker is Frank Mazza and the cellar master is Gary Mosier. The winery operates extensions at 7440 McKnight Road, Pittsburgh, Pennsylvania 15237, and at 2546 West 26th Street, Erie, Pennsylvania 16506.

PENN-SHORE VINEYARDS, INC. / 10225 East Lake Road, North East, Pennsylvania 16428 George W. Sceiford: (814) 725-9422

OPEN MONDAY–SATURDAY 9 A.M.–9 P.M.; CLOSED ELECTION DAY, THANKSGIVING, CHRISTMAS. WT, RS; P (NEARBY).

GR Native and French hybrids **R** Concord Hotel; Delhurst Country Inn; Mums Motel; Bear Lake Inn **HM** The Mums Motel; Ramada Inn; Holiday Inn **HS** Lake Shore Railway Museum; St. Mary's Seminary; Presque Isle State Park.

MISC Penn-Shore has been in operation ever since the passage of Pennsylvania's Limited Winery Act of 1968. These vineyards produce ten Still wines and two Sparkling wines, Champagne and Kir.

THE GREAT LAKES REGION

PRESQUE ISLE WINE CELLARS / 9440 Buffalo Road, North East, Pennsylvania 16428   Marlene or Douglas Moorhead: (814) 725-1314

OPEN TUESDAY–SATURDAY 8 A.M.–5 P.M. WT, RS, P.

GR Riesling; Chardonnay; Aligote; Pinot Gris; Cabernet Sauvignon; Dutchess; De Chaunac; Vidal; Sevyal Blanc; Chambourcin   **R** Delhurst Country Inn; Ricardo's; South Shore Inn; Barnacle Bill's; Concord Hotel   **HM** Mums Motel; Rainbow Motor Court   **HS** Presque Isle State Park; Drake Oil Well.

MISC It was the Moorheads' interest in amateur wine making that led them to plant the first *Vitis vinifera* in this area in 1958. The current winery was added in 1969 after the passage of Pennsylvania's Limited Winery Act, but the Presque Isle winery still specializes in supplying the home wine maker with equipment and quality grapes for home production. The visitor is welcome to tour this small winery.

New York

JOHNSON ESTATE WINERY / Box 52, West Main Road (Route 20), Westfield, New York 14787   William A. Gulvin: (716) 326-2191

OPEN IN WINTER MONDAY–SATURDAY 9 A.M.–5 P.M.; OPEN IN SUMMER DAILY 9 A.M.–5 P.M.; CLOSED FOURTH OF JULY, THANKSGIVING, CHRISTMAS. WT, RS.

GR French-American hybrid; *labrusca*   **R** Bark Grill; Webb's Captain's Table; Good Morning Farm Restaurant   **HM** Webb's Captain's Inn; Inn-at-the-Peak; Colonial Squire; Holiday Inn   **HS** Chautauqua Institute; Barcelona Lighthouse; McClurg Mansion and Museum.

MISC In 1962 Frederick Johnson inherited these vineyards and opened his own winery. As of this writing, owner Johnson and wine maker Gulvin are planning significant expansion.

MERRITT ESTATE WINERY / 2264 King Road, Forestville, New York 14062   (716) 965-4800

OPEN MONDAY–SATURDAY 8:30 A.M.–5:30 P.M.; SUNDAY 1 P.M.–5:30 P.M.; CLOSED HOLIDAYS. WT, RS.

Merritt Estate Winery (cont.)

GR Concord; Niagara, Aurora, Baco; Foch; De Chaunac   **R** White Inn; Vineyard Restaurant; Webb's Captain's Table; Colony House   **HM** Holiday Inn; Vineyard Motel   **HS** Chautauqua Institute; Chautauqua Belle Paddle Wheeler.

MISC When the New York State Liquor Authority issued new regulations under which farms were allowed to operate small wineries (under fifty-thousand-gallon capacity), the Merritt Family entered the wine business. Merritt Estate opened on September 18, 1977, and the family takes great pride in its work. Owners William and Marguerite Merritt comment: "We are a small farm winery, very new, and we specialize in excellent wines and sincere hospitality."

MOGEN DAVID WINE CORPORATION / 85 Bourne Street, Westfield, New York 14787   Joseph Baideme: (716) 326-3151
OPEN MONDAY–FRIDAY BY APPOINTMENT ONLY; CLOSED LEGAL AND JEWISH HOLIDAYS, AND DURING THE MONTH OF JULY. WT UPON REQUEST.

GR Concord; Niagara; Catawba; Fredonia   **R** Mar Mar Restaurant   **HM** Holiday Inn   **HS** Chautauqua Institute Cultural Center.

MISC In 1967 what was then the Westfield Sommer's Food Company was purchased by Mogen David of Chicago (see listing, Great Lakes, Illinois).

Ontario

ANDRE'S WINES, LTD. / South Service Road & Kelson Road, Winona, Ontario, Canada   Carol Wright: (416) 561-1811
OPEN MONDAY–SATURDAY BY APPOINTMENT. WT, RS.

WP A wide range of wines available   **R** Innsville; Casablanca   **HM** Prudhomme's Motel; Casablanca Motel   **HS** Stoney Creek Battlefield; Depot.

MISC Founded in 1961 by Andrew Peller, Andre's now owns six wineries throughout Canada. The winery in Winona was purchased from Beau Chatel in 1970 and offers a wide range of wines, which can be tasted and purchased on the premises.

THE GREAT LAKES REGION

CHARAL WINERY & VINEYARDS, INC. / R.R. #1, Blenheim, Ontario, Canada
Allan Eastman: (519) 676-3012

OPEN MONDAY–SATURDAY 10 A.M.–6 P.M. WT, RS, P.

GR French hybrids; *vinifera; labrusca* **R** O-J's Restaurant; Wheels Motor Inn; Venus de Milo **HM** Wheels Motor Inn; Holiday Inn.

MISC At the Wineries Unlimited Competition in 1977, the Charal Pinot Chardonnay 1975 received a gold medal as the overall highest scorer as a premium Vinifera wine. The first wines were produced here in 1975 and marketed in 1977.

JORDAN WINES / 120 Ridley Road, St. Catherines, Ontario, Canada
Mrs. Micky Crowfoot: (416) 688-2140

OPEN MONDAY–FRIDAY 10 A.M.–3 P.M. WT, RS.

WP Champagne; Crackling, Table, Sparkling, and Dessert wines; Aperitifs **HS** Jordan Historical Museum (open May–October); Niagara Falls; Niagara Scenic Route; Welland Canal.

MISC The original Jordan winery buildings were owned by the Smure Family who, since 1870, had been producing apple and grape products. In 1921 the winery was purchased by Archie J. Haynes who soon changed the name to Jordan Wines. Jordan Wines now owns fifty acres of vineyards and grows approximately fifty-five varieties of grapes which are used purely for experimental purposes. Jordan also contracts almost 25 percent of the twenty-five thousand acres of vineyards on the Niagara peninsula.

New York: The Finger Lakes District

The history of wine growing around New York's Finger Lakes is similar to that of other eastern wine districts. Settlers wished to produce wines as early as the 1600s, but they rejected the native grapes they found on the land, probably because the tastes were unfamiliar, and they imported vine stock from vineyards back home in Europe. You can guess the results—the imported vines could not stand the long, harsh eastern winter, nor the virulent vermin and rots which the native strains could withstand. It was not until the early nineteenth century that native varieties were planted around the Finger Lakes, and a systematic effort at viticulture was again undertaken. The first planting of cultivated native vines was made by the Reverend William Bostwick in 1829; by the 1860s the district was producing fine Champagne and wines. Catawba and Alexander grapes were used, and today the region makes wines from an impressive variety of native and hybrid grapes.

The Finger Lakes region is spectacular country to tour in. The region is noted for its gorges, waterfalls, rapid streams, six hundred miles of Lakes shoreline, rich fertile soil, and acres of vine-covered slopes. Most of the wineries are located near the towns of Hammondsport, Naples, and Penn Yan, about a three-hundred-mile drive from New York City. (Take the New York Thruway to Exit 16, then Route 17 west to Bath, then a right turn on Route 54.) The area is very popular with vacationers and sports enthusiasts, so there are many restaurants and places to stay. These towns are wine towns. There are large and small wineries to visit, most offering tours and tastings. The Champagnes from this district have a particularly distinctive history. By the early 1900s European experts were judging certain Finger Lakes vintages as possibly the greatest Champagne in the western world. That is how Great Western Champagne got its name—it is grown here. In this area you are in for a superb wine-tasting experience.

From one of the motels overlooking Keuka Lake you can take a very pleasant few days getting to know the wines and vineyards of the area. In addition, the Baseball Hall of Fame is nearby at Cooperstown, as is a nineteenth-century farmer's museum and restored village; and, just outside town, James Fenimore Cooper's manuscripts are on view at Fenimore House, along with the State Historical Association's collection of local folk arts and crafts. Watkins Glen is east of Hammondsport on Route 14, and if you miss the auto-racing season there (usually September–October), you can still visit Watkins Glen State Park and its eighteen waterfalls and cataracts. Just south of the park is Corning, home of the Corning Glass Center and the Steuben Glass Factory. There you can watch glass being blown and finished; and perhaps you will want to purchase a piece after your tour. The glass museum here contains over fourteen thousand pieces. At Hammondsport you should visit the Finger Lakes Wine Museum. The museum is maintained by local vintners and displays early wine-making equipment as well as the tools and techniques of the entire process. The museum has a research library and an extraordinary grape library of more than two hundred varieties from all over the world. If you are interested in making your own wine, the Museum Winemaker's Shop can provide you with equipment and instructions, and even with grapes and juice. The museum is open from June 1 through October 31, 9:30 A.M.–4 P.M. on weekdays, and on Sundays from 1 P.M.–4 P.M.

THE BARRY WINE COMPANY, INC. (O-NEH-DA VINEYARDS) / 7107 Vineyard Road, Conesus-on-Hemlock, New York 14435
  T. S. ("Ted") Cribari, Jr., and A. B. ("Skip") Cribari III: (716) 346-2321

TOURS MID-MAY–NOVEMBER MONDAY–SATURDAY 10 A.M.–5 P.M.; SUNDAY NOON–5 P.M.; DECEMBER–MID-MAY SATURDAY AND SUNDAY NOON–5 P.M. WT PLANNED; RS PLANNED; P. OPEN FOR WINE TASTING AND TOURS MOST LEGAL HOLIDAYS NOON–5 P.M.

NEW YORK: THE FINGER LAKES DISTRICT

GR Delaware; Catawba; Niagara; Elvira; Riesling; Aurora
WP Sacramental wines and "Barry" Table wines   **R** Conesus Lake Inn; Beachcomber; Callahan's Truehardt's   **HM** Holiday Inn

MISC Founded in 1872 by Bishop McQuaid of Rochester, New York, Diocese as a source of sacramental wines, the winery and vineyards were leased to the Barry Wine Company in 1968 by the Society of the Divine Word, which had purchased the winery from the Diocese in 1920s. Since then the winery has continued to produce premium New York State altar wines, and has expanded its Table wine business.

BULLY HILL VINEYARDS / Greyton H. Taylor Memorial Drive, Hammondsport, New York 14840
(607) 868-3610 or toll free within New York State: 800-252-5408

OPEN MAY 1–OCTOBER 31 MONDAY–SATURDAY 10 A.M.–4 P.M.; SUNDAYS 1 P.M.–4 P.M. WT, RS, P; D (LUNCHEON).

GR French hybrids   **R** Champagne Country Cafe; Snug Harbor Inn; Gerry's Restaurant; Chateau Dugas   **HM** Hammondsport Motel; Bath Ramada Inn.

MISC In 1970, in order to preserve the Taylor Family heritage, Greyton H. Taylor (the son) and Walter S. Taylor (the grandson) established the Bully Hill Wine Company on the original winery property owned by Walter Taylor from 1880 to 1926. This establishment enabled them to maintain their family's tradition. Walter S. Taylor is now the sole owner of the Bully Hill Wine Company. The tasting room at Bully Hill is situated one thousand feet above Keuka Lake, and Bully Hill also offers the visitor the Greyton H. Taylor Wine Museum, Art Gallery, Winemakers' Shop, and Champagne Country Cafe.

KONSTANTIN D. FRANK AND SONS VINIFERA WINE CELLARS, LTD. / R.D. #2, Hammondsport, New York 14840
Dr. Konstantin D. Frank: (617) 868-4884

OPEN BY APPOINTMENT ONLY. WT, RS.

WP Johannisberg Riesling; Chardonnay; Gewürztraminer; Muscat Ottonel; Cabernet Sauvignon; Pinot Noir; Gamay Beaujolais; Pinot Gris; Sereksia
R M. & R. Restaurant; Gerry's Restaurant; Fred's Chicken House
HM Hammondsport Motel; Bath Ramada Inn   **HS** Greyton H. Taylor Wine Museum; Glenn H. Curtiss Museum.

MISC This seventy-acre vineyard, planted in 1962 by Dr. Frank, includes over sixty grape varieties from all over the world. Dr. Frank, a vintner for sixty-five years, continues to experiment with new *vinifera* vines.

**GLENORA WINE CELLARS / Glenora-on-Seneca, Dundee, New York
14837** Gene Pierce: (607) 243-7600

OPEN MAY 15–OCTOBER 31 MONDAY–SATURDAY 10 A.M.–5 P.M.; SUNDAY 1 P.M.–5 P.M.; OPEN NOVEMBER 1–MAY 14 MONDAY–FRIDAY 9 A.M.–3 P.M.; CLOSED NEW YEAR'S DAY, THANKSGIVING, CHRISTMAS. WT, RS.

GR *Vinifera*; French-American hybrids; *labrusca* **R** Glen Motor Court; Pierce's 1894; Belhurst Castle **HM** Glen Motor Court; Corning Hilton **HS** Blue Hill Wine Museum; Watkins Glen State Park; Corning Glass.

MISC This jointly owned 230-acre vineyard offered its first vintage in 1978.

GOLD SEAL VINEYARDS, INC. / West Lake Road, Hammondsport, New York 14840 Wanda Putnam: (607) 868-3232

OPEN NOVEMBER–APRIL MONDAY–FRIDAY 9 A.M.–11 A.M., 1 P.M.–4 P.M.; MAY, JUNE, SEPTEMBER, OCTOBER MONDAY–SATURDAY 9 A.M.–11 A.M., 1 P.M.–4 P.M.; SUNDAYS NOON–4 P.M.; JULY, AUGUST DAILY 9:30 A.M.–4:30 P.M. WT, RS.

WP Champagne; Vintage, Table, Native, and Dessert wines **R** Chateau Dugas; Lakeside Hotel; Snug Harbor; Village Tavern **HM** Ramada Inn; Hammondsport Motel; Vinehurst **HS** Curtiss Museum; Garrett Memorial Chapel.

MISC Gold Seal Vineyards was founded in 1865, and it was known then as the Urbana Wine Company. Producing under the brand name of Gold Seal, these vineyards became a leader in producing New York State Champagne, and their wines were known throughout the United States. The wine-making department at Gold Seal has been, and still is, under the direction of European-trained wine makers.

HERON HILL VINEYARDS / Route 76, R.D. #2, Hammondsport, New York 14840 Peter Johnstone (607) 868-4241

WINE AND LUNCH BAR OPEN DAILY 10 A.M.–5 P.M.; SUNDAY 1 P.M.–5 P.M. MAY–NOVEMBER; CLOSED THANKSGIVING, CHRISTMAS. TOURS BY APPOINTMENT ONLY. WT, RS.

GR Johannisberg Riesling; Chardonnay; Aurora; Seyval **HS** Curtiss Aviation Museum.

MISC Wine maker Peter Johnstone notes that his winery was bonded in 1977. It is his aim to produce modest quantities of high-quality White wine.

NEW YORK: THE FINGER LAKES DISTRICT

PLEASANT VALLEY WINE COMPANY (GREAT WESTERN) / Old Bath Road, Hammondsport, New York 14840

Patrick S. Natale: (607) 569-2121

GUIDED TOURS ARE CONDUCTED CONTINUOUSLY 8:30 A.M.–11 A.M., 1 P.M.–3:30 P.M. THE YEAR ROUND MONDAY–FRIDAY; JULY 7–OCTOBER 13 TOURS CONDUCTED MONDAY–SATURDAY; CLOSED MAJOR HOLIDAYS AND THE DAY AFTER THANKSGIVING. GROUP TOURS BY RESERVATION ONLY. WT, RS.

WP Champagnes; Solera Sherries and Ports; Vermouths; generic and varietal Table wines.

MISC As the oldest winery in the prestigious Finger Lakes wine region, the Pleasant Valley Wine Company (Great Western) has the distinction of being the first winery to win an award in international competition. Their Champagne was honored in Paris in 1871, and received the highest award given to an American Champagne with the Diploma of Honor in Brussels, 1910. Pleasant Valley also produces Sherries and Ports using the Solera aging system.

ROBIN FILS & CIE, LTD. / School Street & Hewitt Place, Batavia, New York 14020

Ned Cooper: (716) 344-1111

OPEN BY APPOINTMENT ONLY.

WP Table wines; Sparkling and Specialty wines; Vermouth **R** Steak House **HM** Holiday Inn; Treadway Inn; Park Oak Motel **HS** Batavia Land Office; Batavia Downs; Letchworth State Park.

MISC Approximately 500,000 gallons of Sparkling wine and an even greater volume of Still wines are produced at this third-generation family winery.

TAYLOR WINE COMPANY / Old Bath Road, Hammondsport, New York 14840

Faye L. Dowdle: (607) 569-2111

OPEN ALL YEAR MONDAY–FRIDAY 8:30 A.M.–11:30 A.M., 1 P.M.–3:30 P.M. CLOSED MAJOR HOLIDAYS AND DAY AFTER THANKSGIVING. GROUP TOURS BY RESERVATION ONLY. WT, RS, P.

GR Native American; French-American hybrids   **WP** Champagnes; Sherries and Ports; Vermouths; generic Table wines.

MISC Founded in 1880 by Walter Taylor, the Taylor Wine Company is today the largest producer of bottle-fermented Champagne in the United States, and the largest producer of premium wines east of the Rocky Mountains. Until 1961 this was a family-owned-and-operated winery. In 1962 the company "went public," and in 1977 Taylor merged with the Coca-Cola Company of Atlanta, Georgia.

VILLA D'INGIANNI WINERY / 1183 East Keuka Lake Road, R.D. #1, Dundee, New York 14837   Isabelle d'Ingianni: (607) 292-6526

OPEN DAILY 9 A.M.–4 P.M.; CLOSED LEGAL HOLIDAYS. TOURS BY APPOINTMENT ONLY. WT, RS.

GR *Vinifera; labrusca*; French hybrids   **R** Country Host; Gerry's
HM Viking Motel; Country Host.

MISC This new winery, opened in 1972 by Dr. Vincente d'Ingianni, has a 140,000-gallon capacity. The winery and 150 acres of vines overlook Keuka Lake.

WIDMER'S WINE CELLARS, INC. / West Avenue, Naples, New York 14512   Nancy S. Van Orman: (716) 374-6311

OPEN MONDAY–SATURDAY 10 A.M.–3:30 P.M.; SUNDAY NOON–4:30 P.M.; CLOSED LEGAL HOLIDAYS. TOURS FOR GROUPS BY APPOINTMENT ONLY. WT, RS.

GR Niagara; Concord; Catawba; Delaware; Moore's Diamond; Aurora; Foch; Vergennes; Riesling; Elvira, Ives; Cayuga White; Vincent; Isabella; Salem; Baco; Dutchess   **R** Redwood; Bob and Ruth's Vineyard
HM Sheraton Inn; Yodel Inn.

MISC These wine cellars were founded in 1888 by John Jacob Widmer. William Widmer brought old-world methods to the winery which he had studied at the Royal Wine School of Germany. Of special interest at the winery are the Sherry barrels aging on the cellar roof in the Solera system aging process.

The New England States (Connecticut, Massachusetts, Rhode Island, and New Hampshire)

New England has a long history of wine growing. Plantings of traditional European grapes did not at first thrive in the northerly climate, however, and it is only in the last few decades that vintners have concentrated their efforts on the kind of vine most likely to do well in the region—hybrids with a hearty native stock. The earliest plantings actually date back to the mid-seventeenth century, in places like Martha's Vineyard, and in and around Boston. Of course, the story of the Concord grape begins in Concord, Massachusetts, where Ephraim Wales Bull first planted them. Bull was interested in the possibilities of native grapes, and he picked native *labruscas* and sowed their seeds, eventually discovering the Concord stock. When he had observed how well they grew, he began a business in Concords, but by that time other viticulturists had come to the same conclusions as he, and Concord plantings were becoming common both in this region and further west. Near Nathaniel Hawthorne's home in Concord is the Bull house, with what is said to be the original Concord vine, the great size and age of which strengthen the claim.

Vineyards in New England continue to be essentially the same as they have been for centuries, although the product is now better. Most wine making in this area is done at small, personally cared-for vineyards and wineries. There are not many of them, but they are worth searching out. An example is the Prudence Island Vineyard, located on Prudence Island, Rhode Island, in a farmhouse built in 1783. A pleasant ride on a ferry takes you to this lovely island, which has an inn near the spot where Roger Williams used to preach. The family which now runs the vineyard is the seventh generation of owners, and they will give you good tips about local hiking and sailing. New Hampshire contains some vineyards in its White Mountain region, which is famous for camping and mountain scenery. The vineyards are becoming known, too, for their flavorful experiments with French-American hybrid grapes and hybrid crosses. The state of Connecticut will soon add its products to the output of the rest of New England, since the state government is now arranging a licensing system for small vineyards and wineries.

In noting several small establishments at which viniferous products of distinction are being developed and offered for sale, we should not forget to mention the backyard arbor from which generations of New Englanders have always made wines. The love of wine, and the taste for experimenting with it, is older and more cherished in this region than in any other area except, of course, the South.

Connecticut

ERBACHER VINEYARDS / River Road, Roxbury, Connecticut 06783
J. R. McMahan: (203) 354-7598

NOT OPEN TO THE PUBLIC.

GR French-American hybrids; experimental vines.

HAMLET HILL VINEYARD / Brayman Hollow Road, Pomfret, Connecticut 06258   A. W. Loos or William Sitts: (203) 928-5550

AS OF SUMMER, 1981: OPEN FRIDAY AND SATURDAY 1 P.M.–5 P.M.; CLOSED MAJOR HOLIDAYS. WT, RS.

THE NEW ENGLAND STATES   135

GR French hybrids   **R** The Place; Hank's; Bald Hill; The Pantry; Chuck's Steak House; Vernon Stiles   **HM** Kings Inn; Berris Motor Inn   **HS** Sturbridge Village; Mystic Seaport.

MISC The vineyards here were planted in the spring of 1976. Wine will not be available to the public until 1981.

Massachusetts

CHICAMA VINEYARDS / Stoney Hill Road, West Tisbury, Massachusetts 02575   Catherine or George Mathiesen: (617) 693-0309

OPEN MAY 30–SEPTEMBER 30 MONDAY–FRIDAY 1 P.M.–5 P.M.; SATURDAY 11 A.M.–5 P.M.; OTHER TIMES BY APPOINTMENT. RS.

GR Chardonnay; Riesling; Cabernet Sauvignon; Pinot Noir; Gewürztraminer; Gamay; Merlot; Pinot Gris   **HS** Many historical homes on Martha's Vineyard.

MISC This new family-owned-and-tended vineyard marketed its first wines in 1974.

Rhode Island

PRUDENCE ISLAND VINEYARDS / Prudence Island, Rhode Island 02872   William Bacon: (401) 683-2452

OPEN DAILY BY APPOINTMENT 9 A.M.–5 P.M. WT, P.

GR Chardonnay; Gamay; Riesling; Cabernet; Gewürztraminer; Merlot   **R** Prudence Inn (July 1–September 1)   **HM** Prudence Inn   **HS** The site from which Roger Williams preached to the Indians.

Prudence Island Vineyards (cont.)

MISC Seven generations of Mrs. Bacon's family have lived on the farmland on which the Prudence Island Vineyards is situated. The farmhouse here was built in 1783. The first planting of vines occurred in 1973 and an underground winery and storage area is under construction. Retail sales are available at the winery. William Bacon writes: "Prudence is a beautiful, largely undeveloped island, in the center of Narragansett Bay. For those who would like to combine a day of hiking with a visit to the vineyard and the winery, this can be a very rewarding experience. The ferry trip is a treat in itself."

SAKONNET VINEYARDS / West Main Road, Little Compton, Rhode Island   Jim or Lolly Mitchell: (401) 635-4356

OPEN MAY–NOVEMBER WEDNESDAY AND SATURDAY 10 A.M.–5 P.M.; OTHERWISE BY APPOINTMENT. WT, P (LIMITED).

GR Red and White French hybrids; *vinifera*   **R** Le Bistro; La Petite Auberge; La Forge Casino; Stone House Club   **HM** Sheraton 60; Castle Hill Inn.

MISC Sakonnet Vineyards covers thirty-five acres and expects to produce thirty thousand gallons of wine annually (currently at 18,500). This vineyard holds Rhode Island's State License #1, and is at present the largest vineyard and winery in New England. There is a self-guided vine walk, tour of the winery, and a complimentary tasting.

WICKFORD VINEYARDS / Chardonnay, 21 Hamilton–Allenton Road, North Kingstown, Rhode Island 02852   (401) 792-2372

VINEYARDS NOT OPEN TO THE PUBLIC.

GR Chardonnay; Riesling; Merlot; Baco Noir   **R** Red Rooster
HS Gilbert Stuart House; Smith's Castle.

MISC The vineyards are still in the preproduction phase.

New Hampshire

WHITE MOUNTAIN VINEYARDS, INC. / R.F.D. 2, Province Road (Route 107), Laconia, New Hampshire 03246

John J. Canepa: (603) 524-0174

OPEN JUNE 1–SEPTEMBER 1 MONDAY–FRIDAY 9 A.M.–NOON, 1 P.M.–4:30 P.M.; SATURDAYS BY APPOINTMENT. VINEYARDS CLOSED DURING HARVEST SEPTEMBER 1–OCTOBER 15. WT, RS.

GR French and American hybrids; new hybrid crosses **R** Hart's Turkey Farm; B. Mae Denny's Restaurant; Denauw's Seafood; Dandy Gander Restaurant **HM** Margate; Best Western; Christmas Island; Lord Hampshire.

MISC This singular New Hampshire winery now produces a dozen different wines, and has plans for future expansion. The owners, John and Lucy Canepa, plan an annual production of 25,000 cases, and already their wines are available in Massachusetts, Rhode Island, New York, New Jersey, Florida, and Ohio. Their Rosé was a gold-medal winner in the 1977 Wineries Unlimited Competition of Lancaster, Pennsylvania. In August there is the yearly wine festival in Guilford, New Hampshire, featuring many White Mountain recreational facilities and wine tastings.

New York: The Hudson Valley

The Hudson River Valley is established wine country. Seventy miles up the Hudson from New York City is the oldest winery in continuous operation in America; and the first wine making took place in this valley at the time of French settlement of the area in the late seventeenth century. By the early nineteenth century, substantial commercial wine making was taking place as far south on the Hudson as Croton Point, with vineyards planted in native varieties. The Brotherhood Winery (not a monastery) has the largest cellar vaults of any winery in the United States. It is located near Washingtonville, off the New York State Thruway, which runs up the western bank of the Hudson. At the Brotherhood Winery there are also tours and tastings, and arrangements can be made for private parties in the evenings. The samplings, and the vintages available for sale, are primarily native grape products, with a few hybrid varieties. Good champagnes have been produced in this region for more than a hundred years, as have a variety of Fruit wines and Sacramental (especially Kosher) wines.

The Hudson Valley, on both sides of the river, is one of the richest scenic and historic districts in America. Cultural sites, museums, tours, and performances can be found throughout the area, primarily in the summer and autumn months. If you travel north from New York City along the western bank, the New York State Thruway will take you through Bear Mountain State Park, West Point, the Storm King Art Center in Mountainville, Washingington's Headquarters outside Newburgh, and New Paltz's seventeenth-century Huguenot restoration, to the resort area center at Catskill. A drive up Route 9, on the eastern bank of the river, might begin around Tarrytown with visits to Washington Irving's home at Sunnyside (Sleepy Hollow country), or Jay Gould's estate at Lyndhurst, or the restored eighteenth-century Dutch village at Philipsburg Manor. Further north, at Katonah, the Caramoor estate hosts a lively music and arts festival all summer, with an evening concert series that is worth inquiring about (914-232-4206). Boscobel Mansion is outside Garrison, farther up the river. It is an eighteenth-century restoration with evening entertainment in the summer. Poughkeepsie, and Vassar College, are just south of Hyde Park, where you can tour FDR's home and the magnificent Vanderbilt mansion, both on one ticket bought at either place.

The same riverside landscape that surrounds the Vanderbilt house is the site of many acres of vines throughout the valley, with high, peaked hills by the river, and rolling fields. The gravelly soils of the area have always produced interesting vintages, as you will discover when dining in the river towns. Points as far north as Hyde Park lie within a well-planned day's travel from New York City. Although not in the Hudson Valley, the Hargrave Vineyard in Long Island is listed here for the purpose of inclusion, even though there is a significant distance between the Hudson Valley and Long Island.

BENMARL WINE COMPANY / Marlboro, New York 12542
Mark Miller or Eric Miller: (914) 236-7271

OPEN MONDAY–SATURDAY 10 A.M.–5 P.M. BY APPOINTMENT ONLY. WT, RS, P. FEE: VARIABLE.

GR *Vinifera*; European direct producers **R** Ship's Lantern Inn; Culinary Institute of America; De Puy Canal House; Bird and Bottle Inn; Harrald's; Marcel's **HM** Beekman Arms; Broadhead House; Mohonk Mountain House; Holiday Inn; Bird and Bottle Inn; Howard Johnson's **HS** Roosevelt Home; Hyde Park; Vanderbilt Mansion; Ogden Mills Estate; West Point Academy; Vassar College; Rhinebeck Aerodrome; Boscobel.

MISC These vineyards were established in the early 1800s by A. J. Caywood, a pioneer viticultural scientist. From 1957 to 1972 the Miller Family replanted these vineyards as a model for the renaissance in Hudson region viticulture. Visitors have the opportunity to apply for membership in the *Société des Vignerons* (when memberships are open). This is a research and development operation providing access to wines not otherwise available.

NEW YORK: THE HUDSON VALLEY

THE BROTHERHOOD WINERY / 37 North Street, Washingtonville, New York 10992 Anne Farrell Lahey: (914) 496-3661 or 496-9101

OPEN DAILY JUNE–AUGUST 7 DAYS A WEEK 10 A.M.–4 P.M. OFF-SEASON HOURS VARY. WT, RS, P.

WP Dinner, Dessert, Varietal, and Sparkling wines **HM** Howard Johnson's; Ramada Inn; Holiday Inn **HS** West Point Academy; New Windsor Cantonment; Washington's Headquarters; Goshen Hall of Fame.

MISC Founded in 1839 by Jean Jacques, Brotherhood Winery has the distinction of being America's oldest continuously producing winery. Brotherhood has traditionally been a leading producer of Champagnes and of Sacramental wines. The winery prides itself on the high quality of its tours, as Anne Farrell Lahey explains: "The wine-tasting tour is designed to be informative, entertaining, and unhurried. Its success can be judged by the more than five million visitors who have come to Brotherhood in the last fifteen years."

CAGNASSO WINERY / Route 9W North, Marlboro, New York 12542 J. Cagnasso or June Ramey: (914) 236-4630

OPEN 10 A.M.–4:30 P.M. MAY 1 THROUGH OCTOBER DAILY; SAME HOURS SATURDAY AND SUNDAY NOVEMBER–CHRISTMAS AND MONTH OF APRIL, CLOSED EASTER, CHRISTMAS. WT, RS.

GR European-American; *labrusca* **R** Marlo Inn; Ship's Lantern Inn; Coppola's; Mariner's Harbor **HM** Atlas Motor Lodge; Three Penny Inn **HS** Washington's Headquarters; Woodstock Art Colony; Vanderbilt Mansion; Roosevelt Home and Library; Ellenville Ice Caves.

MISC Joseph Cagnasso, who has made wines in both his native Italy and the United States, encourages small, individual tastings and tours. His wines are available only at the winery.

CASCADE MOUNTAIN VINEYARDS / Flint Hill, Amenia, New York 12501 William or Margaret Wetmore: (914) 373-9021

OPEN BY APPOINTMENT ONLY. WT, RS, P.

GR French-American hybrid **HM** Altamont Inn; Interlaken Inn; Old Drover's Inn.

MISC William and Margaret Wetmore planted fourteen acres of French-American hybrids in 1972. During the summer of 1977 the Wetmores' sons built the present winery, and in the fall of that year the Wetmores had their first crush.

CLINTON VINEYARDS / Schultzville Road, Clinton Corners, New York 12514
Ben Feder: (914) 266-5372

OPEN BY APPOINTMENT ONLY. WT, RS (LIMITED).

GR Seyval Blanc; Aurora; Ravat 51; Chardonnay; Riesling **R** Country Tavern; Beekman Arms **HS** Roosevelt Home; Hyde Park; Vanderbilt Mansion.

MISC Bonded in 1977, the Clinton Vineyards' first vintage was a Seyval Blanc, 1977. Frank Prial of the *New York Times* described it on November 29, 1978, as "the best White wine from French-American hybrids made in New York State." On December 1, 1978, it won a gold medal and Best of Category, Eastern Wine Conference (Wineries Unlimited), Lancaster, Pennsylvania.

GREAT RIVER WINERY / Marlboro, New York 12542
(914) 236-4646

TASTING ROOM AND WINE CELLAR OPEN DAILY MAY–DECEMBER 11 A.M.–6 P.M.; OPEN JANUARY–APRIL FRIDAY–SUNDAY 11 A.M.–6 P.M.

WP Aurora; Seyval Blanc; Baco Noir; Vincent Noir; Blanc d'Aurore Champagne.

MISC The Great River Winery was founded as Marlboro Champagne Cellars (see listing) in 1944. The Great River name was adopted to reflect its more recent production of premium vintage wines made from French hybrid grapes. Great River wines are made predominantly from grapes grown on two nearby vineyards which together comprise the largest planting of French hybrid vines in the Hudson wine region. In this gravel soil and stimulating climate these vines produce wines of unique delicacy and character. The Great River Blanc d'Aurore Champagne, which is individually fermented in its original bottle in the true *méthode champenoise*, is made only in limited quantities from each year's vintage.

HARGRAVE VINEYARD / Alvah's Lane, Cutchogue, Long Island, New York 11935
Alexander Hargrave: (516) 734-5158

OPEN WEEKDAYS 10 A.M.–4 P.M. RS.

GR Cabernet Sauvignon; Pinot Noir; Chardonnay; Sauvignon Blanc; Merlot; Semillon **R** Moveable Feast; La Gazelle **HM** Beachcomber **HS** Flea Market; Antique Car Show; The Old House.

MISC The town of Cutchogue was the location of the eighteenth-century vineyards of Moses Fournier. The Hargrave Vineyard, planted in 1973, is now the only vineyard on Long Island. It offered its first release in 1977.

NEW YORK: THE HUDSON VALLEY

HUDSON VALLEY WINE COMPANY, INC. / Blue Point Road, Highland, New York 12528   John Lahey: (914) 691-7296 or (212) 594-5394

OPEN JANUARY–MARCH SATURDAY AND SUNDAY 11 A.M.–4 P.M.; APRIL–NOVEMBER DAILY 11 A.M.–3 P.M.; JULY–AUGUST 11 A.M.–4P.M.; DECEMBER SATURDAY AND SUNDAY: "CHRISTMAS IN THE COUNTRY" TOURS; CLOSED THANKSGIVING, CHRISTMAS. FEE: WEEKDAYS $2.00 PER ADULT, CHILDREN FREE; INCLUDES TOUR, TASTING, PARKING, BREAD, FRUIT, AND CHEESE. WEEKEND OPEN HOUSE 11 A.M.–4:30 P.M. FEE: $4.00 PER ADULT, MINORS $1.00; INCLUDES ABOVE PLUS WINE GLASS AND HORSE-DRAWN HAY OR SLEIGH RIDES. CALL FOR SCHEDULES. WT, RS, P; D BY RESERVATION.

GR Chelois; Iona; Delaware; Catawba; Baco Noir; Niagara; Concord; Marechal Foch   **R** Ship's Lantern Inn; Mariner's Harbor; Coppola's; Spat's Fireside Inn; Treasure Chest; DuPuy Canal House; Quilted Giraffe
HM Rocking Horse Ranch; Hyde Park Motel; Red Bull Inn; Mohonk Mountain House; Holiday Inn; Poughkeepsie Motor Hotel
HS Minnewaska State Park; Edwin Ulrich Museum; Kingston State House; Cary Arboretum.

MISC The Bolognesi Family brought their home wine-making talents with them when they emigrated from Italy. This Hudson Valley winery was started in 1907, and it is built in the style of an old European village, with stone buildings and courtyards, surmounted by a clock tower. There are 320 acres under vine along more than a mile of scenic Hudson River property. Aside from wine tours with tastings of Labrusca, and French hybrid Table and Sparkling wines, the Hudson Valley Winery plays host to weekend wine festivals once a month starting the first weekend in April; picnic weekends, July and August; harvest festivals, October; cross-country skiing (conditions permitting) during the winter. In September there is the annual North American Grape Stomping Championship.

KEDEM ROYAL WINERY / Dock Road, Milton, New York 12547
                                            (914) 795-2240

OPEN SUNDAY–FRIDAY 10 A.M.–5 P.M.; CLOSED JEWISH HOLIDAYS. WT, RS, P. FEE: $1.00 PARKING CHARGE ON SUNDAYS AND LEGAL HOLIDAYS.

GR Concord; French hybrids; De Chaunac   **HS** West Point Academy; Hyde Park.

MISC Wine maker Ernest Herzog is a sixth-generation vintner from Czechoslovakia whose family wines have won gold medals in Vienna and Budapest. In 1948 the Herzog family left occupied Czechoslovakia for New York's Lower East Side. Bringing the Herzog know-how with them, they purchased these Milton vineyards in 1967. Today the winery produces over one million gallons of wine annually, of twenty-six different flavors from very dry to sweet wines, and Fruit wines, Honey wines, Sangria, and others. Tours of the winery are offered, including a film on vineyard production and bottling. Wine tasting and occasional enological seminars are also available.

MARLBORO CHAMPAGNE CELLARS / 104 Western Avenue, Marlboro, New York 12542 (914) 236-4440

OPEN MONDAY–FRIDAY 8 A.M.–4 P.M. WT, RS.

WP New York State Champagnes and wines.

MISC These wine cellars, founded in 1944, produce Champagnes under the Chaumont and Marlboro labels. Chaumont and Marlboro Champagnes are bottle-fermented in the traditional *méthode champenoise*.

NORTHEAST VINEYARD / Millerton, New York 12546
Dr. George Green: (518) 789-3645

OPEN BY APPOINTMENT ONLY. RS.

GR French hybrid   **R** The Verandah.

MISC The Northeast Vineyards, planted in 1971, specialize in Foch and Aurora vines.

SCHAPIRO WINE COMPANY / 126 Rivington Street, New York, New York 10002 (212) 674-4404

OPEN SUNDAYS 10 A.M.–6 P.M. WT, RS.

WP Concord; Cream Plum; Cream Honey Mead; Cream Almondetta; Fruit and Berry wines; Table and Sparkling wines.

MISC Located in the heart of Manhattan, the Schapiro Wine Company, of course, is not truly in the Hudson Valley region. Established in 1899 by Samuel Schapiro, this company is a landmark location, and the only winery in New York City today. This is the reason I have included it. Using grapes from upstate New York, Schapiro's presents a full line of strictly Kosher wines, and welcomes Sunday visitors.

The Mid-Atlantic Region (Pennsylvania, New Jersey, and Maryland)

Pennsylvania's long history in the making of wine has been somewhat obscured by the state's liquor monopoly as well as by the slow start made by vineyards after the repeal of Prohibition. But, in fact, the cultivation of wine grapes in Pennsylvania began as early as the mid-seventeenth century, with efforts by Swedish settlers outside of Philadelphia. William Penn, in the 1680s, planted Spanish and French vines, and one of his staff started a vineyard that shipped wines through the 1750s. During the Revolution, John Alexander discovered the Alexander grape, and soon the first good native American wine-making stock to be cultivated spread from Alexander's home to vineyards as far distant as Indiana, via the Ohio River.

The late nineteenth century was a busy time for Pennsylvania vintners as well, for besides the establishment of vineyards around Lake Erie, a vital wine industry grew up in Allegheny County, near Pittsburgh. Growing there was paced by Father Rapp's Harmony Society, and by the middle of the nineteenth century Allegheny County was the largest wine-producing area in the state. Wine making also went on in the Pennsylvania Dutch district around York and Lancaster. In fact, Pennsylvanians claim that the first plantings in the Finger Lake district of New York came from York, Pennsylvania, cuttings of 1829.

The modern era of wine making in Pennsylvania began in the southeastern part of the state, when the First French hybrids were planted in Chester County. In the ensuing years, Pennsylvania State University established several experimental vineyards across the state, and now maintains an experimental winery to test the thirty-odd varieties of grapes that grow in Pennsylvania. If you are touring vineyards in Pennsylvania, you have several opportunities for interesting scenic and historical sightseeing. The New Hope area is close to such Revolutionary War sites as Valley Forge and Washington's Crossing. The Lancaster and York area wineries are located in the Pennsylvania Dutch country, a rich farmland known for folk art and for the quaint ways of the Amish community, unchanged for centuries.

New Jersey is the home of various kinds of vineyards on its muddy soils—from the large Renault wineries on the southern shore at Egg Harbor City to small family vineyards in the historic townships around Princeton. The House of Renault makes its guests welcome with, among other things, a Blueberry Champagne. All of these vineyards are accessible from either New York City or Philadelphia.

Wine was being grown in Maryland as early as 1662, when Lord Baltimore planted grapes at St. Mary's. In the early nineteenth century Major John Adlum discovered the Catawba grape for wine growing outside Washington, D.C., and grew it successfully in Maryland, calling his vintage a "Tokay." Thomas Jefferson wrote that he had enjoyed the bottle which Major Adlum had sent him as a gift.

Present-day wine making in Maryland, as in many other eastern states, owes much to the efforts of Philip M. Wagner of Baltimore. In the 1930s, after Prohibition, Wagner planted a variety of hybrids from France. He popularized these hearty hybrids throughout the East, and soon found himself as busy in the nursery business as he was at wine making. The other vineyards of Maryland where his progeny now grow are situated in lovely country, and the families that run them are proud to help you appreciate an ongoing tradition of fine Maryland wines.

Pennsylvania

ADAMS COUNTY WINERY / R.D. #1, Orrtanna, Pennsylvania 17353
Ronald F. Cooper: (717) 334-4631
OPEN MONDAY–SATURDAY 9 A.M.–6 P.M.; CLOSED ELECTION DAY. WT, RS, P.

WP Pinot Chardonnay; Riesling; Gewürztraminer; Seyval Blanc; Vidal Blanc; Foch; Chelois; De Chaunac; Leon Millot; Bordeaux **HS** Gettysburg; Pennsylvania Dutch country.

MISC Started in 1974, this winery and vineyard specializes in European-style wines. The winery shows its own film on wine making at special times.

BUCKINGHAM VALLEY VINEYARDS / Box 371, Route 413, Buckingham, Pennsylvania 18912 **Kathy Forest: (215) 794-7188**
OPEN TUESDAY–FRIDAY 2 P.M.–7 P.M.; SATURDAY 10 A.M.–6 P.M. WT, RS.

GR French-American hybrids **R** Stone Manor Inn; Peddlers Village; Chez Odette; Tom Moore's Tavern; Conti's Inn **HM** Holiday Inn **HS** Town of New Hope; Washington's Crossing; Valley Forge.

MISC This small family-owned-and-operated vineyard, planted in 1966, has a total annual production of ten thousand gallons of Red, White, Rosé, and Apple wines.

BUCKS COUNTRY VINEYARDS / Route 202, New Hope, Pennsylvania 18938 **Arthur Gerold: (215) 794-7449**
OPEN MONDAY–SATURDAY 11 A.M.–6 P.M.: CLOSED NEW YEAR'S DAY, ELECTION DAY, THANKSGIVING, CHRISTMAS. APPOINTMENT NECESSARY FOR GROUPS OVER TEN. WT, RS.

GR Baco Noir; Foch; Chelois; Seyval Blanc; Aurora **R** Hacienda Inn; Broadmoor; Hotel du Village; La Bonne Auberge; Le Bistro; Black Bass Hotel **HM** Lambertville House; Holiday Inn; 1740 House; Hotel du Village; Motel-in-the-Woods **HS** Washington's Crossing State Park and Museum; Mercer Museum; Moravian Tile Works; Shrine of Czestohova.

MISC Although the Bucks Country Vineyards were licensed as recently as 1973, they have already won numerous medals and ribbons at the Pennsylvania Wine Competition in Erie, at the Maîtres des Tastevins in Washington, D.C., and at the Host Farm Wine Competition. Annual production here is now 46,000 gallons, all of which is marketed at retail prices at the winery, or is

Bucks Country Vienyards (cont.)

available in approximately seventy restaurants in eastern Pennsylvania. The winery features seasonal festivals, including a May wine festival, a Sangria festival with Spanish dancers in June, an Italian festival with foods and dancers in July, a Polish wine festival in August, and, rounding out the season, an Oktoberfest. The wine and fashion museum at the winery features many wine artifacts, a history of the vine, and original theatrical costumes of many famous Broadway stars.

CHÂTEAU PIATT WINERY, INC. / 108 North Ninth Street, Allentown, Pennsylvania 18102 (215) 821-1415

OPEN TUESDAY–SATURDAY 10 A.M.–5:30 P.M. WT, RS.

WP Niagara; various hybrids; Concord.

CONESTOGA VINEYARDS / Flowing Springs Road, Birchrunville, Pennsylvania 19421 (215) 822-9024

OPEN SATURDAY NOON–5 P.M. SPECIAL TOURS ARE AVAILABLE FOR GROUPS, BY APPOINTMENT ONLY, DURING THE WEEK. WT, RS; P (LIMITED).

GR French hybrids; *vinifera*   **R** Coventry Forge Inn; Kimberton Inn; French Creek Falls Hotel   **HM** Holiday Inn; Coventry Forge Inn   **HS** Valley Forge Campgrounds and Museum; St. Peters Village; Hopewell Village.

MISC Owner David Charles Fondots notes that these vineyards were first planted in 1957, making the Conestoga Vineyards the oldest commercial vineyards and winery in Pennsylvania. Mr. Fondots has recently opened his second wine-tasting and sales shop. It is located at 2323 Lincoln Highway East, Lancaster, Pennsylvania.

DUTCH COUNTRY WINE CELLAR / R.D. #1, Box 15, Lewhartsville, Pennsylvania 19534   Helen Tenaglia: (215) 756-6061

OPEN MONDAY–FRIDAY 1 P.M.–5 P.M.; SATURDAY 9 A.M.–6 P.M.; CLOSED LEGAL HOLIDAYS. WT, RS, P.

GR French hybrids; *labrusca*   **R** Regal Restaurant; Glockenspiel; Haag's   **HM** Regal Motel   **HS** Hawk Mountain Bird Sanctuary.

MISC The Tenaglia Family notes that their vineyard, planted in 1970, was originally a hobby. The wine cellar soon became a greater challenge. The

THE MID-ATLANTIC REGION 149

Tenaglias now produce two thousand gallons of wine each year. All the work is done by hand within a converted Pennsylvania Dutch-style barn featuring Hex Sign decorations.

LEMBO VINEYARDS / 34 Valley Street, Lewistown, Pennsylvania 17044
Joseph Lembo: (717) 248-4078

OPEN MONDAY-SATURDAY 9 A.M.-9 P.M. WT, RS.

GR French hybrids   **R** Tony's Cottage Inn   **HM** Holiday Inn; Hotel Lewistown; Green Cables Motor Inn; Steven's Motel; Elks Club.

MISC The vineyards were planted in 1972 and the winery was established in 1976.

NISSLEY VINEYARDS / R.D. #1, Box 92-B, Bainbridge, Pennsylvania 17502
(717) 426-3514

OPEN MONDAY-SATURDAY NOON-5 P.M.; CLOSED NEW YEAR'S DAY, FOURTH OF JULY, THANKSGIVING, DECEMBER 24-25. WT, RS. WINERY TOUR FEE: $1.00.

WP French hybrid, Varietal, and Table wines   **R** Railroad House; Mr. Lacy's; Colonial Inn; Hillcrest; Three Center Square Inn   **HM** Blue Note Motor Inn; Quality Inn   **HS** Haldeman Mansion; Bainbridge Blacksmith Shop; Donegal Mills Colonial Plantation; Pennsylvania Dutch Country.

MISC Set among rolling farmland hills, the Nissley Vineyards winery adjoins an eighteenth-century stone mill.

PEQUEA VALLEY VINEYARD AND WINERY / Box 332, Willow Street, Pennsylvania 17584
H. Peterman Wood: (717) 464-3721

OPEN MONDAY-SATURDAY 9 A.M.-5 P.M.; CLOSED NEW YEAR'S DAY, ELECTION DAY, THANKSGIVING, CHRISTMAS. WT, RS, P.

GR French hybrids   **R** Willow Valley Farms; Lemon Tree; Brunswick Hotel   **HM** Willow Valley Farms; Strasburg Motel.

MISC This winery, built in a converted dairy farm, made its first sale in 1973.

WILMONT WINERY / Mill Road, R.D. #2, Schwenksville, Pennsylvania 19473
Frank Wilmer: (215) 287-8089

OPEN TUESDAY-FRIDAY NOON-5 P.M.; SATURDAY 11 A.M.-5 P.M.: CLOSED ELECTION DAY. WT, RS; P. INVITED (BUT NO FACILITIES).

Wilmont Winery (cont.)

GR Chardonnay; Cabernet Sauvignon; Riesling; Seyval Blanc; Foch; Chelois; Baco; Villard Blanc **R** Fagleysville Hotel; Lakeside Inn **HM** Holiday Inn; Sheraton; Stouffer's **HS** Valley Forge Park.

MISC These vineyards were planted in 1971 in the midst of apple and peach orchards. The winery, licensed in 1976, has an annual capacity of six thousand gallons. Visitors may view the wine-making process from the retail sales area within the winery building.

New Jersey

ANTUZZI'S WINERY / Bridgeboro–Moorestown Road, Delran, New Jersey 08075 (609) 764-1075
OPEN MONDAY–WEDNESDAY NOON–6 P.M.; THURSDAY AND FRIDAY NOON–8 P.M.; SATURDAY 10 A.M.–8 P.M. WT, RS.

WP Baco Noir; Seyval Blanc; Ravat 51; Chablis; Rosé; Pink Catawba; White Catawba; Delaware; Niagara; Sweet Claret; Blackberry; Strawberry; Raspberry; Apple d'Apple; Cherry Royale **R** Pirate's Inn; Pavin's Stoned Crab; Rancocas Inn **HM** Holiday Inn; TraveLodge; Mt. Laurel Hilton **HS** Rancocas Woods.

MISC Owner and wine maker Matthew J. Antuzzi is naturally proud that his Baco Noir was selected at a blind wine tasting to travel to Rome in honor of the canonization of Bishop John Neuman. His Seyval Blanc won a silver medal in a Washington, D.C. competition. Opened in 1974, the winery now has an annual capacity of twenty thousand gallons.

BALIC WINERY / Route 40, Mays Landing, New Jersey 08330
Savo Balic: (609) 625-2166
OPEN MONDAY–SATURDAY 9 A.M.–6 P.M. WT, RS.

GR Fredonia; Sheridan; Delaware; Ives; Noah; Catawba; French hybrids.

THE MID-ATLANTIC REGION 151

GROSS HIGHLAND WINERY / 212 Jim Leeds Road, Absecon, New Jersey 08201   (609) 652-1187

OPEN MONDAY–SATURDAY 9 A.M.–6 P.M.; CLOSED NEW YEAR'S DAY, EASTER, FOURTH OF JULY, THANKSGIVING, CHRISTMAS. GROUP TOURS BY APPOINTMENT ONLY. WT, RS.

GR Native and hybrid   **HS** Batsto Restoration; Brigantine Wildlife Refuge.

MISC This family-owned winery, opened in 1934, offers tasting bars, tours of the wine cellars and vineyards, and a narrated slide presentation.

RENAULT WINERY / R.D. 3, Box 21B, Egg Harbor City, New Jersey 08215   Barbara Muller: (609) 965-2111

OPEN MONDAY–SATURDAY 10 A.M.–5 P.M.; CLOSED NEW YEAR'S DAY, GOOD FRIDAY, THANKSGIVING, CHRISTMAS. FEE: $1.00 PER ADULT. WT, RS.

GR Noah; Ives; Catawba; Norton; Fredonia; Missouri Riesling; Elvira; Delawares; Niagara; Aurora Blanc; De Chàunac; Baco Noir; Seyval Blanc; Vidal; Foch; Chaucellor Noir; Chardonnay   **HS** Smithville Village; Batsto Village; Renault Vineyard.

MISC The House of Renault, with its 1,400 acres and 300,000-gallon winery, was founded by master vintner Louis Nicholas Renault more than a century ago. He come to the United States representing the Champagne House of the Duke of Montebello of Reims, France. In 1864 he purchased land for the present Renault vineyard. By 1870 he had introduced his New Jersey Champagne. The Winery has been in operation ever since, even during Prohibition when it sold Renault Wine Tonic nationwide, making it the oldest winery in continuous operation in the country. The Winery is unique in its production of Noah Blanc, Blueberry Champagne, Blueberry Duck, and "Pink Lady." Visitors are welcome to this official New Jersey Historical Site which houses one of the finest collections of antique wineglasses, dating back to the thirteenth century. Tours are available. The gift shop offers a selection of miscellaneous items and Renault Champagnes and wines.

TOMASELLO WINERY / 225 North White Horse Pike, Hammonton, New Jersey 08037   (609) 561-0567

OPEN MONDAY–SATURDAY 9 A.M.–8 P.M.; CLOSED HOLIDAYS. WT, RS.

GR Native American and French hybrids   **HS** Batsto Historical Park.

MISC The Tomasello Winery encompasses approximately one hundred acres and was founded in 1933.

Maryland

BERRYWINE PLANTATION / Route 4, Box 247, Mount Airy, Maryland 21771   Lucille Aellen: (301) 662-8687 or 829-2297

OPEN MAY, JUNE, SEPTEMBER, OCTOBER SATURDAY AND SUNDAY 1 P.M.–DUSK; JULY AND AUGUST OPEN DAILY EXCEPT WEDNESDAY 1 P.M.–DUSK. FEE FOR WINE AND CHEESE TASTING: $1.60. WT, RS, P.

GR French hybrids   **R** Mealy's; Quail Ridge Inn; Red Horse Steak House; Mountain View Inn   **HM** Holiday Inn; The Strawberry Inn; Red Horse Motor Lodge.

MISC Owners Jack and Lucille Aellen inherited their knowledge and appreciation of wines from their parents in Brooklyn, New York. Jack's father made excellent light German wines while Lucille became expert in her parents' heavier Italian wines. The Aellen, Jr., Family has now settled at Berrywine, a farm of 230 acres, producing hay, Simmental beef, and black crossbred sheep; they also have six acres of grapevines. The Aellens reflect their family devotion to quality by saying, "When our wine is ready for bottling, all the family participates. One fills the bottles, another will insert the cork, another will apply the decorative cap, while another will apply the label and return the bottle to its case. Lucille keeps us supplied with foods and chips. When you visit us on our tour you may find that one of us will be your guide, while another will serve you wine and cheese, and still another will help you select the wine you want to take home with you."

BOORDY VINEYARDS / Box 38, Riderwood, Maryland 21139
     J. and P. Wagner: (301) 823-4624

OPEN THURSDAYS AND FRIDAYS BY APPOINTMENT ONLY.

GR French hybrids.

MISC The Boordy Vineyards ask not to be listed as a "tourist attraction," although the Wagners are always willing to meet with serious wine enthusiasts or professionals. Philip Wagner, vintner and wine author, was instrumental in introducing French hybrid grapes commercially to the United States.

BYRD VINEYARDS / Church Hill Road, Myersville, Maryland 21773
     (301) 293-1110

OPEN JUNE–NOVEMBER SATURDAY AND SUNDAY 1 P.M.–6 P.M. GROUPS BY APPOINTMENT. WT, RS. $1.00 PER ADULT, DEDUCTIBLE FROM PURCHASE.

THE MID-ATLANTIC REGION 153

GR Chardonnay; Sauvignon Blanc; Cabernet Sauvignon; Muscat Canelli; Gewürztraminer; Seyval Blanc; Vidal Blanc **R** Cozy Restaurant; Red Horse; Mealy's; Dandee Restaurant; Tortuga; South Mountain Inn **HM** Red Horse Inn; Holiday Inn; Sheraton; Ramada Inn **HS** New Market, Maryland; Francis Scott Key Memorial.

MISC Although founded in 1972, Byrd Vineyards has already enjoyed a colorful history. This family-owned-and-operated winery was trapped during the change in winery laws in Maryland, and for a period Byrd was the "world's most exclusive winery," licensed to make wines, but forbidden to sell those wines to anyone, or even to consume them privately. New legislation has been passed, and although Byrd is perhaps no longer as exclusive, it is now allowed to bring its wines to market. Because this is a family-operated winery, tours are fitted into a tight schedule, when possible. The Byrds sell amateur and commercial winery equipment. Also available, upon request, is membership in their Vintage Reserve Society.

MONTBRAY WINE CELLARS, LTD. / 818 Silver Run Valley, Westminster, Maryland 21157   G. Hamilton Mowbray: (301) 346-7878

OPEN SUNDAY 2 P.M.–6 P.M. WT. FEE: $2.00 PER ADULT.

GR French hybrid; *vinifera* **R** At the Winery **HS** Union Mills Homestead; Gettysburg Battlefields and Museum; Carroll County Farm Museum and Historical Society.

MISC Montbray Wine Cellars was established in 1966.

PROVENZA VINEYARDS / 805 Greenbridge Road, Brookeville, Maryland
    Barbara Provenza: (301) 277-2447 or 774-2310

OPEN BY APPOINTMENT ONLY. WT, D.

GR Vidal 256; Villard Blanc 12375; Seyval Blanc 5276; Cascade 13053; Chamboucin 26205; Chancellor 7053 **R** Silo Inn; Olney Ale House.

MISC Visitors to the Provenza Vineyards are given a thorough tour of their wine-making operation, followed by a tasting of all available wines, usually served with cheeses and bread. The fifteen acres of French hybrids were planted here between 1970 and 1973, and the modern winery, designed by Dr. Provenza, was completed in 1974. All of the vines at Provenza, and much of the winery equipment, were purchased through Philip Wagner of Boordy Vineyards (see listing). Provenza Vineyards wines are 100 percent estate bottled.

The Southern States (Virginia, North Carolina, and South Carolina)

Scuppernong grapes. Other kinds of vines now grow in the South, largely due to the resurgent interest in wine growing and the experimentation with French hybrid stock that southern agriculturists and growers have undertaken in the years since Prohibition. But the southern wine flavor is that of the Scuppernong grape, named from the Algonquin *ascopo* ("sweetbay"), indigenous to the states south of the Mason-Dixon line. This is the spreading vine behind countless southern houses; it is extolled or mentioned in all forms of American literature. It was discovered by Sir Walter Raleigh on Roanoke Island (legend has it), and its taste is considered pleasant, interesting, and distinctive. Scuppernong wines come from Muscadine grapes, the various kinds of which cover the majority of vine acreage in the South.

The Muscadine family of grapes can be any of a variety of hues, and besides producing Scuppernong wines, they can also be eaten fresh or preserved. In those areas which comprise the coastal lands along the Atlantic, Muscadine grapes are the only variety that has thrived in numbers. French hybrids are safer inland in the South, largely because of the storm weather provided by the Atlantic well into the Carolinas.

America came to know the taste of Scuppernong wines under the name "Virginia Dare." Captain Paul Garrett, who became extremely wealthy making and selling a Scuppernong beverage, hit on the Virginia Dare trademark because that was the name of the first baby born in Virginia colony; and he kept the Muscadine taste dominant in his blend throughout its history. After Prohibition he had trouble finding enough Muscadine grapes, so he toured the South and the halls of government during the 1930s to promote the growth of Scuppernong vines. But many areas in the South remained dry through those years, and so it is only in the past few decades that, in the case of the growing Georgia wine areas, for instance, large-scale Muscadine plantings have occurred.

The historical background of the southern wine district is as old as any lore in America. What Raleigh's scouts saw at Roanoke in the late sixteenth century was grapes—everywhere. Verrazano reported them to the French in 1524; Washington grew grapes at Mount Vernon; Jefferson experimented with European soils on his farms. The great native grapes of the South are not now restricted to Muscadine. Isabellas and Nortons, among others, come from southern regions. These are wine districts with great traditions, and they share a serious and innovative approach to the growing of southern natives and European hybrids.

Virginia

BARBOURSVILLE CORPORATION / P.O. Box F, Barboursville, Virginia 22923   Gabriele Rausse: (703) 832-3824

OPEN MONDAY–SATURDAY 7:30 A.M.–6:30 P.M.

GR Chardonnay; Pinot Noir; White Riesling; Gewürztraminer; Merlot; Cabernet Franc; Malvasia Bianca; Cabernet Sauvignon; Gamay Beaujolais   **R** Boar's Head Inn   **HM** Econo-Lodge; Ramada Inn; Holiday Inn   **HS** University of Virginia; Barboursville Ruins; Monticello.

MISC Thomas Jefferson designed the mansion in which James Barbour lived while he served as governor of Barboursville. The remains of this are now known as the Barboursville Ruins. The Barboursville Corporation, founded in 1976, plans to produce its own wines by 1981.

THE SOUTHERN STATES

FARFELU VINEYARD / Rappahannock County, Flint Hill, Virginia 22627
C. J. Raney: (703) 364-2930

OPEN SATURDAY AND SUNDAY 9 A.M.–5 P.M. GROUPS BY APPOINTMENT. WT, RS.

GR French-American hybrids; *vinifera*   **R** The Depot   **HM** Quality Court   **HS** Bull Run Battleground.

MISC The winery at Farfelu is housed within a 150-year-old barn that has been remodeled with minimal exterior change. Annual production here is only twelve hundred gallons of Seyval Blanc and Dry Red. The vineyards were planted in 1967.

GUILFORD FARM VINEYARD / Route 2, Box 51, Stanley, Virginia 22851
John Gerba: Contact in writing; vineyards not open to public until 1980 or 1981.

OPEN BY APPOINTMENT ONLY.

GR French-American hybrids; experimental *vinifera*   **R** Brown's; Big Meadows Lodge; Skyland Lodge; Mimslyn   **HM** Big Meadows Lodge; Skyland Lodge; Mimslyn Cardinal   **HS** Skyland Drive; Shenandoah National Park; Luray Caverns; George Washington National Forest.

MISC Owner and wine maker John Gerba established this small vineyard in 1972 and plans first production by 1980 or 1981. Interim harvests are going into experimental wines and will be sold to home wine makers.

LA ABRA FARM AND WINERY / Route 1, Box 139, Lovingston, Virginia 29949
A. C. Weed II: (804) 263-5392

OPEN WEEKENDS 1 P.M.–5 P.M. AND BY APPOINTMENT AT OTHER TIMES; CLOSED EASTER, THANKSGIVING, CHRISTMAS. WT, RS. P.

GR French hybrids; also apples and peaches   **R** Wintergreen Resort; Rutledge Inn   **HM** Village Motel; Wintergreen resort   **HS** Monticello; University of Virginia; Madison and Monroe Homes; Natural Bridge.

MISC La Abra Farm and Winery produces Fruit wines as well as French hybrid wines. The first grapes were planted in 1974, and the winery was bonded in 1976.

MEREDYTH VINEYARDS (at Stirling Farm) / Route 628, Middleburg, Virginia 22117
Mr. or Mrs. Archie M. Smith: (703) 687-6612 or 687-6277

OPEN BY APPOINTMENT ONLY 10 A.M.–4 P.M.; CLOSED EASTER, CHRISTMAS WT, RS.

Meredyth Vineyards (cont.)

GR French-American hybrids; *vinifera* **HM** Holiday Inn; Sheraton Hotel; Marriott Hotel; Stouffer's **HS** Manassas Battlefield; Stratford (home of Robert E. Lee); Mount Vernon, Sully Plantation; Oatlands Plantation.

MISC Wine maker Archie Smith notes that Meredyth Vineyards is staffed by industrious young people who will try to welcome visitors despite their busy schedule. The vineyards here were planted in 1972, and Meredyth has been producing wines since 1976. Today there are some thirty-five acres under vines, and there are plans for expansion.

PATOWMACK VINEYARD / 209 Seneca Road, Great Falls, Virginia 22066   Bill Garrett: (703) 450-4160

NOT OPEN TO THE PUBLIC.

GR Pinot Chardonnay; French hybrids.

SCHIEHALLION VINEYARDS / Route 2, Box 29-B, Keysville, Virginia 23947   Pat Barber

OPEN BY WRITTEN APPOINTMENT.

GR French hybrids; *vinifera*; experimentals **R** Sheldon's; Sol Brown's; Hotel Weyanoke; Royal Gardens **HM** Sheldon's Motel; Hotel Weyanoke **HS** Patrick Henry's grave and last home at Red Hill; Appomattox Civil War Battleground.

MISC The Southern Piedmont region is famous for its tobacco growing. Schiehallion Vineyards is currently experimenting with over thirty varieties of grapes to establish those plants and growing techniques best suited for wine production in such a region.

VINIFERA WINE GROWERS ASSOCIATION / Experimental Vineyard, The Plains, Virginia 22171
R. de Treville Lawrence, Sr: (703) 754-8564

OPEN BY APPOINTMENT ONLY. WT LIMITED; P.

GR *Vinifera* **R** Red Fox Tavern; L'Auberge Inn; Le Rat **HM** Ramada Inn; Holiday Inn.

MISC These vineyards are the headquarters of the Vinifera Wine Growers Association, the publishers of *Jefferson and Wine* and Vinifera Wine Growers (quarterly) *Journal*.

North Carolina

WINE CELLARS, INC. / Route 2, Edenton, North Carolina 27932
Frank L. Williams: (919) 482-4295

OPEN MONDAY–FRIDAY 9 A.M.–5 P.M.; SATURDAY 10 A.M.–4 P.M.; CLOSED FOURTH OF JULY, LABOR DAY, THANKSGIVING, CHRISTMAS. WT, RS, P.

GR Scuppernong; Carlos **R** Soundview Restaurant; Boswell; Coach House Inn **HM** Coach House Inn; Holiday Inn **HS** Tryon Palace; Historic Edenton Tour; Somerset Plantation at Pettigrew State Park; Hope Plantation; Wright Memorial; Fort Raleigh.

MISC Although a census taken in 1810 by James Blount indicated that 1,368 gallons of wine had been made that year in Washington County, the Wine Cellars of Edenton is the only incorporated winery in North Carolina today. The vines here were planted in 1971, and now there are some eighty acres under cultivation. The winery features a tasting room within a renovated "red barn" and offers tastings of dry Scuppernong wines as well as White Table wines, Rosé, and Sweet Red wines. Production is beginning and will include a native Champagne bottled under the Deerfield label.

South Carolina

OAKVIEW PLANTATIONS / P.O. Box 342, Woodruff, South Carolina 29388
R. L. Leizear or Barbara Judd: (803) 476-3112

OPEN MONDAY–FRIDAY 8 A.M.–4:30 P.M.; CLOSED MAJOR HOLIDAYS. WT, RS.

GR Muscadine; Scuppernong; Catawba **R** Townhouse **HM** Holiday Inn; Sheraton Motor Inn.

MISC Although the winery is a new attraction at Oakview, grapes have been cultivated on 371 acres of this plantation for over ten years. Here grapes have been stemmed, crushed, frozen, and stored in grape pulp containers with a total capacity of 1.8 million pounds. According to USDA reports, Oakview now has fifteen to forty percent of the frozen grape pulp in America in its freezers. The winery now has a fully automatic production facility for a line of Still wines, with an annual capacity of over one million cases. Future plans include a fully automated line of Champagnes and Sparkling wines.

TRULUCK WINERY / Route 3, Drawer 1265, Lake City, South Carolina 29560 (803) 389-3400 or 394-2318

OPEN FOR TOURS AND TASTINGS TUESDAY–SATURDAY 10:30 A.M.–5:30 P.M. WT, RS, P.

GR *Vinifera*; French hybrids   **HM** Lake City Motel; Holiday Inn.

MISC This newly established winery, done in the Old World style, represents the aspirations of Dr. and Mrs. James Truluck upon their return from two years in the USAF Dental Corps in France's Loire Valley. The winery, completed in 1976, has many characteristics of those found in France. It is a small, family operation, run by two generations of Trulucks. Here, Dry and Sweet Table wines are produced from Villard Blanc and Chambourcin stock. Special classes on viticulture are offered from September through October. Future plans are for a château on the premises with a European restaurant manned by a French chef. All this will overlook Lake Truluck, located in the midst of the vineyards.

The Upper Mississippi Region (Iowa and Missouri)

The Midwestern states that lie along the Mississippi River grow more native grape varieties than do the other wine-producing districts of North America, and residents of these states share a greatly increased taste for drinking wines, as well as a resurgent interest in growing wine grapes. Prohibition effectively wiped out Midwestern vineyards, although there had once been wineries in forty-eight of Missouri's counties, where nineteenth-century wine traditions flourished. In those days St. Louis was America's principal center of wine study. It was supplied with the fruits of wine production, then going on in both the Mississippi and Ohio river valleys, and with products from vineyards planted in the Ozark plateau country of Arkansas. The people of St. Louis also enjoyed vintages made by the Amana Colonies of Iowa.

Now many traditional Midwest vineyard districts are producing wines again, and, although Prohibition took a heavy toll in the Midwest, there is increasing experimentation in raising French hybrids and in cultivating the numerous native varieties as well. In Missouri, for example, you might plan a trip from St. Louis to vineyards in the Gasconade and Bourbeuse river districts, to the historic towns of Augusta and St. Charles, and to others farther afield in the southeastern quadrant of this state. Touring these districts affords glimpses of a highly original wine heritage as well as tastes of the conscientious and innovative efforts of present-day winegrowers—a dedicated band of enthusiasts who are benefiting from the improved wine technology of modern American viticulture.

Iowa

ACKERMAN WINERY / Highway 6 at Highway 220, South Amana, Iowa 52334 (319) 622-3379

OPEN MONDAY–SATURDAY 9A.M.–6 P.M.; CLOSED CHRISTMAS. WT, RS, D.

WP Grape; Rhubarb; Strawberry; Cherry; Dandelion; Apple; Plum; Blueberry; Elderberry **R** Colony Market Place **HM** Die Heimat Motel; Holiday Inn **HS** Numerous historical sites, museums, and craft exhibits within the Amana Colony area.

MISC Opened in Amana in 1956, the Ackerman Winery offers a gift shop featuring German steins and glassware as well as local Amana craftware.

CHRISTINA WINE CELLARS / 123 A Street, McGregor, Iowa 52157
Christine Lawlor: (319) 873-3321

OPEN APRIL–DECEMBER MONDAY–SATURDAY 10 A.M.–6 P.M. WT, RS, P.

WP Montmorency Cherry; Apricot; Apple; Concord; Catawba **R** White Springs; Red Cedar Inn; Pink Elephant; Village Traveller; Scenic Hotel Coffeeshop **HM** Holiday Shores; Frontier Inn; Pink Elephant Motel
HS The towns of McGregor, Iowa, and Prairie du Chien, Wisconsin, offer numerous historical sites.

MISC Wine maker Christine Lawlor has, among other distinctions, that of being one of the first women enologists to graduate from the University of California at Fresno's enology department. With the assistance of her family, the first wines at Christina were sold in the spring of 1976. The active Lawlor Family now plans to open a second winery in La Crosse, Wisconsin.

COLONY WINE AND CHEESE HAUS / Interstate 80, Amana Exit #225, Iowa

OPEN DAILY IN SUMMER 8 A.M.–9 P.M.; WINTER 9 A.M.–5 P.M. NO WINE SALES ON SUNDAY. WT, RS, P, D.

GR Concord; Fredonia; Beta **R** Colony House **HM** Best Western; Holiday Inn **HS** Numerous historical sites, museums, and craft exhibits within the Amana Colony area.

MISC There is a cheese and wine tasting room at Colony Wine and Cheese Haus, where wines are still made in accordance with traditional methods.

THE UPPER MISSISSIPPI REGION

EHRLE BROTHERS / Homestead, Iowa 52236
(319) 622-3241

OPEN MONDAY–SATURDAY 10 A.M.–6 P.M. WT, RS.

WP Rhubarb; Grape   **HM** Die Heimat; Amana Holiday Inn
HS Numerous historical sites, museums, and craft exhibits within the Amana Colony area.

MISC The Ehrle Brothers have been in business since 1941.

SANDSTONE WINERY / Box 7, Amana, Iowa 52203
(319) 622-3081

OPEN MONDAY– SATURDAY 9 A.M.–8 P.M.; CLOSED NEW YEAR'S DAY, THANKSGIVING, CHRISTMAS. WT, RS.

WP Concord Grape; Niagara; Rhubarb; Cherry   **R** Colony Inn; Ox Yoke Inn; Ronneburg Restaurant   **HM** Die Heimat; Holiday Inn; Colony Haus Motel   **HS** Numerous historical sites, museums, and craft exhibits within the Amana Colony area.

MISC This winery, started in 1960, is located in one of the oldest of Amana's houses, built of native sandstone in 1855. As with all Amana wineries, this is strictly a family business, with an annual capacity of approximately ten thousand gallons. All wines are sold locally through the winery.

VILLAGE WINERY / Amana, Iowa 52203
(319) 622-3448

OPEN MONDAY–SATURDAY 9 A.M.–5 P.M. WT, RS.

WP Fruit wines   **R** Colony Inn; Ox Yoke Inn; Ronneburg   **HS** Numerous historical sites, museums, and craft exhibits within the Amana Colony area.

Missouri

BARDENHEIER WINE CELLARS / 1019 Skinker Parkway, St. Louis, Missouri 63112 (314) 862-1400

OPEN MONDAY–FRIDAY 9:30 A.M.–3 P.M.; CLOSED LEGAL HOLIDAYS. WT, RS; D (BY RESERVATION FOR GROUPS OF THIRTY-FIVE OR MORE).

WP Dry Red and White; Mellow and Sweet Red and White; Catawba; Niagara; French hybrid   **R** Cheshire Inn; Quality Inn; Town Hall; Chase Park Plaza; Bel Air West; Stan Musial's   **HM** Cheshire Inn; Quality Inn; Chase Park Plaza; Bel Air West   **HS** St. Louis Art Museum; Forest Park; St. Louis Zoo; Municipal Opera; Jefferson Memorial; Museum of Science and Natural History; Shaw's Botanical Garden; Historical Society of Missouri.

MISC This winery was established in 1873 by John Bardenheier and is still operated by members of the Bardenheier Family.

BOWMAN WINE CELLARS / 500 Welt Street, Weston, Missouri 64098
(816) 386-5235

OPEN MONDAY–SATURDAY 9 A.M.–6 P.M.; SUNDAY 11:30 A.M.–5 P.M. CLOSED NEW YEAR'S DAY, THANKSGIVING, CHRISTMAS. WT, RS, P, D.

WP Table and Sparkling wines   **R** America Bowman Keeping Room (at the winery); Platte Purchase Barn   **HM** Air Port Inn; Hilton Inn; Ramada Inn; Holiday Inn   **HS** Contact Pat O'Malley. Historic Weston Committee, Box 97, Weston, Missouri 64098, 816-386-5235, for tours of historic attractions, including the Bowman Wine Cellars, and for special dining information.

MISC In 1842 the Weston Brewing Company built four extensive caves for the manufacture of beer. Today three of these are used for the aging of Bowman wines. At the winery is a period restaurant, the America Bowman Keeping Room (reservations in advance). Here, hostesses in mid-nineteenth-century costume serve visitors at handcrafted tables set with pewter service and hand-thrown pottery. Cheese and wine tastings are also available for groups.

GREEN VALLEY VINEYARDS / State Road D, R.R., Portland, Missouri 65067   N. A. Lamb: (314) 676-5771

OPEN WEDNESDAY–SATURDAY AND BY APPOINTMENT. WT, RS; P NEARBY.

GR French hybrids; *vinifera*   **HS** Westminster College; Churchill Memorial; St. Mary of Aldermanbury Church.

MISC The vines at Green Valley were planted during the 1960s and the first wines were offered in 1973. The emphasis here is placed on limited wine production using European methods.

KRUGER'S WINERY AND VINEYARDS / Route 1, Nelson, Missouri
65347   Harold Kruger: (816) 784-2325

OPEN SATURDAY AFTERNOONS. WT, RS.

GR *Labrusca*; Muscadine; French hybrids   **R** Black Sheep Inn; Marshall Inn; Sarley's; Pioneer Restaurant; Old Tavern   **HM** Marshall Inn
HS National Monument to George Caleb Bingham; Daniel Boone, Dr. John Sappington, and William Becknell Memorabilia.

MISC Some of the vinestock at Kruger's Vineyards was brought to this country from Germany in 1860 by ancestors of the current owner and wine maker. The winery is a new addition, built in 1976, and operated along traditional German lines.

MISSOURI STATE FRUIT EXPERIMENT STATION / Southwest Missouri State University, Mountain Grove, Missouri 65711
                                                                 Kenneth W. Hanson: (417) 926-4105

OPEN MONDAY–FRIDAY 8 A.M.–NOON, 1 P.M.–5 P.M.; CLOSED HOLIDAYS.

MISC Since the Missouri State Fruit Experiment Station was established in 1900, fruit-breeding work has been carried on in an effort to create better cultivars for fruits adaptable to Missouri soil and climate. Although all types of grapes are grown here, these are not grown for commercial or enological purposes.

MONTELLE VINEYARDS / Route 1, Box 94, Augusta, Missouri 63332
                                                        Nissel or Clayton Byers: (314) 228-4464

OPEN TUESDAY–SATURDAY 11 A.M.–6 P.M. WT, RS, P; D BY ARRANGEMENT.

GR *Vinifera-Riparia* hybrids; American varietals   **R** Ponticello's; Embers; Jim and Charlie's; Femme Osage; Tony's; Anthony's   **HM** Suburban Motel   **HS** Daniel Boone's Home; Gateway Arch; St. Louis Zoo; St. Louis Art Museum; Jefferson Memorial (Missouri Historical Society Museum); Municipal Opera (summer); Missouri Botanical Gardens; extensive cultural and athletic events in St. Louis.

MISC Husband and wife Clayton and Nissel Byers founded Montelle Vineyards in 1970, after ten years of study of wines and vineyards. From seventy-

Montelle Vineyards (cont.)

three varieties first planted, they have winnowed to forty-one. The emphasis is on fermenting skins, even the whites briefly for body and character, blending compatible varietals for balance, and the use of the pressure formula of Bordeaux's Lafite for softness. (The first vintage, 1973, is sold exclusively through Anthony's Restaurant in St. Louis.) Montelle wines are available under the Miraclair and River Country labels, primarily full bodied Dry Whites, with one unique proprietary Aperitif Red, Wenceslaus, with twenty-nine fruit-herb components. Byers heads Missouri's wine education and research committee and loves to discuss wine with visitors.

MOUNT PLEASANT VINEYARDS / #1 High Street, Augusta, Missouri 63332   (314) 228-4419

OPEN MONDAY–SATURDAY 10 A.M.–5:30 P.M.; CLOSED NEW YEAR'S DAY, THANKSGIVING, CHRISTMAS. WT, RS, P.

GR French-American hybrids; American varietals   **R** Femme Osage; St. Charles Vintage House and Wine Garden.

MISC In the period before Prohibition, when Missouri ranked as the second-largest wine-producing state in the nation, there were no fewer than eleven wineries in the town of Augusta. The largest and most famous of these was Mount Pleasant, whose wines, grown from vineyards planted by German immigrants in 1860, had won medals at the Columbian Exposition of 1893 and the St. Louis World's Fair of 1904. It is still the policy of vintner Lucian W. Dressel to ferment and bottle his wines by traditional European methods. Under the Emigré label, Mount Pleasant bottles Red, White, and Rosé wines. Also available are special wines such as Cynthianna, Münch, Missouri Riesling, and Red and White Concord.

OZARK VINEYARDS / Chestnutridge, Missouri 65630
(417) 587-3555

OPEN MONDAY–SATURDAY 9 A.M.–9 P.M.; CLOSED THANKSGIVING, CHRISTMAS. WT, RS, P.

GR French-American hybrids.

MISC This small vineyard was established in 1976.

PEACEFUL BEND VINEYARDS / Route 2, Box 131, Steelville, Missouri 65565   Dr. A. N. Arneson or N. Arne Arneson: (314) 775-2572

OPEN MONDAY–SATURDAY 9 A.M.–5 P.M.: CLOSED HOLIDAYS. WT, RS.

GR French-American hybrids **HS** Museum and original Iron Works at Meramec Springs; Meramec State Park.

MISC These vineyards were originally planted in 1965, and the grapes were sold to home wine makers and local vintners. Bonded in 1972, Peaceful Bend had its first bottling that same year.

ROSATI WINERY / Route 1, Box 55, St. James, Missouri 65559
R. H. Ashby: (314) 265-8629

OPEN MONDAY–SATURDAY 9 A.M.–6 P.M.; CLOSED NEW YEAR'S DAY, CHRISTMAS. TOURS DURING SUMMER MONTHS. WT, RS, P.

GR Concord; Elvira; Missouri Riesling; French hybrids **HS** Maramel Springs; Old Iron Works; Museum.

MISC The Rosati Winery is named after an Italian–American farming community founded in the 1890s. Wine has been made here since the early 1900s, although the current winery was built in 1972, after having been destroyed by a fire. Rosati features a self-guided tour as well as a guided tour during the summer months.

ST. JAMES WINERY / Route 2, Box 98, St. James, Missouri 65559
(314) 265-7912

OPEN MONDAY–SATURDAY 8 A.M.–DARK; CLOSED ELECTION DAY. WT, RS, P.

GR Native American; French hybrids; *vinifera* **HM** Forest City Motel **HS** Meramec State Park.

MISC Owners Pat and Jim Hofherr founded their winery in 1970.

STOLTZ VINEYARDS AND WINERY / R.F.D. #2, St. James, Missouri 65559
W. B. Stoltz: (314) 265-8873

WINERY CLOSED TEMPORARILY. CHECK FOR REOPENING IN 1980.

GR American; American hybrids; French hybrids **R** Manor Inn; Zeno's; Houston House; Chub and Jo Restaurant; Nickerson Farms; Sirloin Stockade **HM** Zeno's; Wayfarer; Forest City; Manor Inn; Holiday Inn; Howard Johnson's **HS** Meramec Iron Works; Meramec Springs.

MISC The Stoltz Winery was started in 1967 as an adjunct to the vineyards, which have 132 acres under vines. Over one hundred varieties of grapes are grown here, and there is a two-acre lot set aside as a demonstration plot for the University of Missouri.

STONE HILL WINE COMPANY / Route 1, Box 26, Hermann, Missouri
65041   L. James Held: (314) 486-2221

OPEN JANUARY–FEBRUARY MONDAY–SATURDAY 8 A.M.–5 P.M.; MARCH–DECEMBER MONDAY–SATURDAY 8 A.M.–5 P.M. AND SUNDAY NOON–5 P.M.; CLOSED NEW YEAR'S DAY, THANKSGIVING, CHRISTMAS. WT, RS.

GR Catawba; Concord; Missouri Riesling; Norton Seedling; Niagara; French-American hybrids.   **R** Rockhouse Cafeteria; Central Hotel; Hillebrands Coffee Shop; Calico Cupboard   **HM** Germann Haus; Riverview; Hermann Motel.

MISC At the turn of the century the town of Hermann was said to export more wines than any other town in the United States. The many gold medals won at that time by Stone Hill Wine Company attest to the quality of these wines. The company offers tours through their historic wine cellars, and features a museum with an antique wine display and a gift shop.

ZIEGLER WINERY / Star Route, Box 20-C, Cuba, Missouri 65453
                    Laurence and Charlene Ziegler: (314) 885-7496

OPEN SATURDAY–MONDAY 9 A.M.–4 P.M. WT, RS, P.

GR Concord   **R** Dutch Barn   **HM** Midway Motel; Wagon Wheel
HS Onondoga and Meramec Caverns; Meramec Springs.

MISC In 1966 Laurence Ziegler purchased this farm, which included twenty-five acres of grapevines. The winery now has a five-thousand-gallon fermenting capacity, as well as wooden barrel storage for 2,500 gallons.

Addendum

Arkansas

CENTER RIDGE WINERY / Star Route, Springfield, Arkansas 72157
(501) 354-5881
Partners: James Rossi, Dolores Rossi

COWIE WINE CELLARS / P.O. Box 284, Paris, Arkansas 72855
(501) 963-3990
Owner and General Manager: Robert G. Cowie

DE SALVO'S WINERY / Route 1, Box 45, Center Ridge, Arkansas 72027
Owner: A. B. De Salvo

FREYALDENHOVEN'S WINERY / Route 1, Highway 64 West, Morrilton, Arkansas 72110 (501) 354-4241
Owner: Wm. D. Freyaldenhoven

HECKMANN'S WINERY / Route 1, Box 148, Harrisburg, Arkansas 72432
(501) 578-5541

MOUNT BETHEL WINERY / U.S. Highway 64, Altus, Arkansas 72821
(501) 468-2444
Owner: Eugene J. Post

NEIL'S WINERY, INC. / Route 1, Springdale, Arkansas 72764
(501) 361-2954
President: Emmett V. Neil

POST WINERY INC. / Route 1, Box 1, Altus, Arkansas 72821
(501) 468-2741
President: Mathew J. Post

HENRY J. SAX / Altus, Arkansas 72821

WIEDERKEHR WINE CELLARS, INC. / Wiederkehr Village, R.R. 1, Box 14, Altus, Arkansas 72821 (501) 468-2611 or 468-3611
Owners: The Wiederkehr Family

California

AHERN WINERY / 715 Arroyo Avenue, San Fernando, California 91340
Mailing address: 14612 Saticoy Street, Van Nuys, California 91405
(213) 989-3898
Owner and Wine maker: James P. Ahern

ALATERA VINEYARDS / 5225 Highway 29, Napa, California 94558
(707) 944-2914
President and Wine maker: Bruce M. Newlan

AMADOR WINERY / Highway 49, P.O. Box 166, Amador City, California 95601
(209) 267-5320
Owner: Leland F. Merrill

BALLARD CANYON CORP. WINERY / 1825 Ballard Canyon Road, Solvang, California 93463 (805) 688-7585
Owner: Ballard Canyon Corp., President: Gene Hallock

B & R VINEYARDS, INC. / 4350 North Monterey Highway, Box 247, Gilroy, California 95020 (408) 842-5649
Owned by B & R Vineyards, Inc., President: John P. Rapazzini

BANDIERA WINES / 155 Cherry Creek Road, Cloverdale, California 95425
(707) 894-2352
Owner and Wine maker: Marc H. Black

BEAULIEU VINEYARD / 1960 St. Helena Highway, Rutherford, California 94573
(707) 963-3671
President: L. F. Knowles, Jr.

BELIZ-DE LOACH VINEYARDS / 3349 Industrial Drive, Santa Rosa, California 95401 (707) 528-1599 or 544-7504
Officer: Berle Beliz

BELLA NAPOLI WINERY / J21128 South Austin Road, Manteca, California 95336
(209) 599-3885
Owner: Estate of Tony R. Hat; Lucas G. Hat, executor

BORRA'S CELLAR / 1301 East Armstrong Road, Lodi, California 95240
(209) 368-5082
Owners: Stephen J. Borra and Beverly V. Borra

BROOKSIDE ENTERPRISES, INC. (A subsidiary of Beatrice Foods, Inc.) /
9900 Guasti Road, Guasti, California 91743 / (714) 983-2787
President: Rene Biane

ADDENDUM

BUEHLER VINEYARDS / 820 Greenfield Road, St. Helena, California
94574 (707) 963-2155
Owner: John P. Buehler, Jr.

DAVIS BYNUM WINERY / 8075 Westside Road, Healdsburg, California
95448 (707) 433-5852
Owner: Davis Bynum Winery Inc., President: Davis Bynum

CADENASSO WINERY / Box 22, Fairfield, California 94533
(707) 425-5845
President: Frank G. Cadenasso

CALIFORNIA GROWERS WINERY INC. / Executive Office: 30 Hotaling
Place, San Francisco, California 94111 (415) 398-1111
President: Robert Setrakian

THE CANNERY WINE CELLAR / 2801 Leavenworth Street, San
Francisco, California 94133

CARMEL BAY WINERY / P.O. Box 2496, Carmel, California 93921
Owners: Fred Crummey and Bob Eyerman

CHAPPELLET WINERY / 1581 Sage Canyon Road, St. Helena, California
94574 Owner: Donn Chappellet

CHÂTEAU MONTELENA WINERY / 1429 Tubbs Lane, Calistoga,
California 94515 (707) 942-5105
Owner and General Partner: James L. Barrett

LOUIS CHERPIN WINERY / 15567 Valley Boulevard, Fontana, California
92335 (714) 822-4103
Owner: Eugene L. Cherpin

CHISPA CELLARS / 425 Main Street, Murphys, California 95247
(209) 728-3492
Owners: Robert J. Bliss and James L. Riggs

COARSEGOLD WINE CELLAR / On southeast side of state Highway 41,
about three miles south of Post Office, Coarsegold, California 93614
Owner: H. A. Button

CONROTTO WINERY / 1690 Hecker Pass Highway, Gilroy, California
95020 (408) 842-3053
Owner: Anselmo Conrotto

CONSOLIDATED DISTRIBUTION SERVICES, INC. / 33369 Transit
Avenue, Union City, California 94587 (415) 489-6140
President: Creed H. Jenkins

CONSUMNES RIVER VINEYARD / 1917 P Street, Sacramento, California
95814 (209) 245-6208
Owner: E. C. Story

CUCAMONGA VINEYARD COMPANY (California Bonded Winery Number 1) / 10013 8th Street, P.O. Box 607, Cucamonga, California 91730
(714) 987-1716
Owner: Pierre Biane

CUCAMONGA VINTNERS / 10277 Foothill Boulevard, P. O. Box 696, Cucamonga, California 91730 (714) 987-2509
President: Arthur Accomazzo

D'AGOSTINI WINERY / Route 2, Box 19, Shenandoah Road, Plymouth, California 95669 (209) 245-6612
General Manager: Armenio D'Agostini

DEHLINGER WINERY / 6300 Guerneville Road, Sebastopol, California
95472 (707) 823-2378
Co-owner: Tom Dehlinger

DELANO GROWERS CO-OPERATIVE WINERY / Route 1, Box 283, Delano, California 93215 (805) 725-3255
President: John Buksa

DEL REY WINERY / 5427 East Central Avenue, Fresno, California 93725
(209) 264-2901
General Manager: Russell J. Murray

DIGARDI WINERY / P.O. Box 88, 3785 Pacheco Boulevard, Martinez, California 94553 (415) 228-2638
Owner: Francis J. Digardi

DURNEY VINEYARD / P. O. Box 1146, Carmel Valley, California 93924
(408) 625-1561
Owner: W. W. Durney

ENZ VINEYARDS / 1781 Limekiln Road, Hollister, California 95023
(408) 637-3956
Owners: Robert W. and Susan W. Enz

ESTRELLA RIVER WINERY / Shandon Star Route, Highway 46, Paso Robles, California 93446 (805) 238-6300
President: Clifford R. Giacobine

FARNESI WINERY / 2426 Almond Avenue, Sanger, California 93657
(209) 875-3004
Owner: Danny C. Farnesi

ADDENDUM

FERREIRA WINES, INC. / 5990 Wine Road, Newcastle, California 95658
(916) 885-9234
President: Alex Ferreira

FIRPO WINERY / Oakley, California 94561   Owner: Julius Firpo

JAMES FRASINETTI AND SONS / P.O. Box 28213, Sacramento,
California 95828   (916) 383-2444
Owners: Gary and Howard Frasinetti

FREDSON WINERY / 18521 Redwood Highway, Geyserville, California 95441
Winery: 1960 Dry Creek Road, Healdsburg, California   (707) 433-1290
Owners: Leonard, Jane, and Josephine Fredson

FRETTER WINE CELLARS / 804 Camelia Street, Berkeley, California
94710   (415) 525-1762
Owner and Wine maker: Travis D. Fretter

GAUTHIER & CLEVENGER LTD. (Also doing business as Vikings Four) / 448 Alisal Road, Solvang, California 93463
Owners: Gauthier & Clevenger Ltd., General Manager: Don Gauthier

PETER AND HARRY GIRETTI / 791 5th Street, Gilroy, California 95020
(408) 842-3857
Owners: Peter and Harry Giretti

GREEN & RED VINEYARD / 3208 Chiles Pope Valley Road, St. Helena,
California 94574   Owner: Jay Heminway

GRGICH HILLS CELLAR / 1829 St. Helena Highway, Rutherford,
California 94573   (707) 963-2784
Owners: Austin Hills and Miljenko Grgich

GUILD WINERIES AND DISTILLERIES / 500 Sansome Street, San
Francisco, California 94111   (415) 391-1100
Chairman of the Board: Hubert Mettler

J. J. HARASZTHY & SON / 14301 Arnold Drive, Glen Ellen, California
95442   (707) 996-3040
Owners: Jan and Vallejo Haraszthy

HARBOR WINERY / 610 Harbor Boulevard, West Sacramento, California
Owner and Wine maker: Charles H. Myers

HEUBLEIN, INC. (Consumer Products Division) / 12467 Baseline,
Etiwanda, California 91739   (714) 987-1751
Owner: Heublein, Inc.; General Manager: R. E. Browne

HORIZON WINERY / 2594 Athena Court, Santa Rosa, California 95401
(707) 544-2961
Owner: Paul D. Gardner

JADE MOUNTAIN WINERY / 1335 Hiatt Road, Cloverdale, California 95425
(707) 894-5579
Owner: Dr. Douglass Sebastian Cartwright

JEKEL VINEYARDS / 10920 Ventura Boulevard, Studio City, California 91604
(213) 769-6414
Winery: 40155 Walnut Avenue, Greenfield, California 93927
Co-owner and General Manager: William D. Jekel

KONOCTI CELLARS (Also doing business as Lake County Vintners) / P.O. Box 925, Kelseyville, California 95451
(707) 279-1712
Owner: Lake County Vintners, President: Walter Lyon

HANNS KORNELL CHAMPAGNE CELLARS / Larkmead Lane, P.O. Box 249, St. Helena, California 94574
(707) 963-2334
President: Hanns J. Kornell

CHARLES KRUG WINERY / C. Mondavi & Sons, P.O. Box 191, St. Helena, California 94574
(707) 963-2761
Product Manager: Peter Mondavi

LA MONT WINERY, INC. / P.O. Box 428, DiGiorgio, California 93217
(805) 845-2231 or 854-2116
Owner: John LaBatt Ltd., Canada, President: Craig D. Crisman

LANDIS VINEYARD / 2068 E. Clayton Avenue, Fresno, California 93725

LA PURISIMA WINERY / 715-A Sunnyvale–Saratoga Road, Sunnyvale, California 94087
(408) 738-1011
General Partner: Doug Watson

LIBERTY WINERY, INC. / 6055 E. Acampo Road, Acampo, California 95220
(209) 368-6646
Owner and President: Herbert Buck

LIVE OAKS WINERY / 3875 Hecker Pass Highway, Gilroy, California 95020
(408) 842-2401
Owner: Peter Scagliotti

LODI VINTNERS, LTD. / 3750 E. Woodbridge Road, Acampo, California 95220
(209) 368-5338
Owner: Lodi Vintners Ltd., Board Chairman: George Bagdasarian

LONG VINEYARDS / P. O. Box 50, St. Helena, California 94574
(707) 963-2496
Owners: Robert and Zelma Long

ADDENDUM

LOS ALAMOS WINERY / P. O. Box 5, Los Alamos, California 93440
Telephone: Dial "0" and ask for Los Alamos (805) 2391
Owner: Samuel D. Hale, Jr.

LUCAS HOME WINE/D. LUCAS VINEYARD / 18196 N. Davis, Lodi, California 95240 (209) 368-2006
Owner, Vineyard Manager, and Wine maker: David Lucas

MARINE WHOLESALE & WAREHOUSE COMPANY / 301 West "B" Street, Wilmington, California 90744

MARKHAM WINERY / 2812 St. Helena Highway N., St. Helena, California 94574 (707) 963-9577
Owner: Markham Advertising Co., Inc., Board Chairman, and President: H. Bruce Markham

S. MARTINELLI & CO. / P. O. Box 549, 227 E. Beach Street, Watsonville, California 95076 (408) 724-1126
Partner: S. C. Martinelli

MATANZAS CREEK WINERY / 6097 Bennett Valley Road, Santa Rosa, California 95404 (707) 542-8242
Owners: David A. and Sandra S. Steiner

GIUSEPPE MAZZONI / Route A, Box 47, Cloverdale, California 95425
(707) 857-3691
Owners: Fred G. Mazzoni and James Mazzoni

MONT LA SALLE VINEYARDS / P. O. Box 420, Napa, California 94558
(707) 226-5566
Owner: Mont La Salle, President: U. R. Portillo

MOUNT RENAISSANCE VINEYARD & WINERY / Box 575, Oregon House, California 95962 (916) 692-1653
President: Karl Werner

MT. VEEDER WINERY AND VINEYARDS / 1999 Mt. Veeder Road, Napa, California 94558 (707) 224-4039
Owners: Michael A. Bernstein and Arlene Bernstein

NAPA VALLEY COOPERATIVE WINERY / P. O. Box 272, St. Helena, California 94574 (707) 963-2335
President: Virgil Galleron

NAPA VINTNERS / 1721 C Action Avenue, Napa, California 94558
Mailing address: P. O. Box 2502, Napa, California 94558
Owner: N. V. Wines Inc., President: Donald C. Ross

NAPA WINE CELLARS / 1247 Walnut Street, Napa, California 94558

NIEBAUM-COPPOLA ESTATES / 1460 Niebaum Lane, Rutherford,
California 94573 (707) 963-9435
Owner: Francis Ford Coppola

NOBLE VINEYARDS, INC. / P. O. Box 31, Kerman, California 93630
(209) 846-7361 or 846-9303
President: Agustin Huneeus

A. NONINI WINERY / 2640 N. Dickenson Avenue, Fresno, California
93711 (209) 264-7857
Partners: Reno A., Gildo J., and Geno A. Nonini

NORDMAN OF CALIFORNIA / 4836 E. Olive Avenue, Fresno, California
93727 (209) 638-9923
President: James W. Hansen

OAK GLEN WINERY / P. O. Box 381, Yucaipa, California 92399
(714) 797-3724
Co-owner and Wine maker: Charles B. Colby

OPICI WINERY, INC. / Highland & Hermosa Avenue, Alta Loma,
California 91701 (714) 987-2710
President: Mary Opici Nimmergut

PASTORI WINERY / 23189 Geyserville Avenue, Cloverdale, California
95425 (707) 857-3418
Owner: Frank Pastori

ROBERT PECOTA WINERY / 3299 Bennett Lane, Calistoga, California
94515 (707) 942-4627
Owner: Robert Pecota

PELLEGRINI BROS. WINES, INC. / P. O. Box 2386, South San
Francisco, California 94080 (415) 761-2811
President: Vincenzo Pellegrini

PERELLI-MINETTI & SONS / P. O. Box 818, Delano, California 93215
(805) 792-3162
President: Jerry K. Stanners

PIRRONE WINE CELLARS / P. O. Box 15, Salida, California 95368
(209) 545-0704
Owner: Alfred F. Pirone

POCAI & SONS / Route 1, Box 231, Calistoga, California 94515
(707) 942-4572

PRESTIGE VINEYARDS / 48980 Seminole Drive, Cabazon, California
92230 Owners: Paul E. Hadley and Peggy R. Hadley

ADDENDUM

REGE WINE CO. / 26700 Dutcher Creek Road, Cloverdale, California 95425
(707) 894-2953
Owner: Eugene Rege

RIO VISTA WINERY, INC. / P. O. Box R., Woodbridge, California 95258
(209) 369-1096
President: Morris Weiss

RITCHIE CREEK VINEYARDS (Also doing business as RICHARD MINOR WINERY) / 4024 Spring Mountain Road, St. Helena, California 94574
(707) 963-4661
Owner: R. P. Minor

RIVER RUN VINTNERS / 65 Rogge Lane, Watsonville, California 95076
(408) 722-7520
Co-owner and Wine maker: William Hangen

RME, INC. / 5950 E. Woodbridge Road, Acampo, California 95220
(209) 369-5861

ROSENBLUM CELLARS / 1775 16th Street, Oakland, California 94612
(415) 834-6067
Owner: K. M. Rosenblum

ROUND HILL CELLARS / 1097 Lodi Lane, St. Helena, California 94574
(707) 963-2228
President and General Manager: Charles A. Abela

RUBIDOUX WINERY / 3477 Arlington Avenue, Riverside, California 92506
(714) 686-5771
Owner: Joe Tavaglione

CHANNING RUDD CELLARS / 2157 Clinton Avenue, Alameda, California 94501
(415) 523-1544
Owner: Channing Rudd

ST. HELENA WINE CO. / 3027 Silverado Trail, St. Helena, California 94574
(707) 963-7108
President: Daniel Duckhorn

SAN ANTONIO WINERY, INC. / 737 Lamar Street, Los Angeles, California 90031
(213) 223-1401
President: Steve Riboli

SAN BENITO VINEYARDS, INC. / 6757 Pacheco Pass Highway, Hollister, California 95023
(408) 637-3992
President and Wine maker: Dr. Rodney W. Ballard

SAN JOAQUIN WAREHOUSING CO. / 6161 E. Highway 12, P. O. Box 728, Lodi, California 95240
(209) 369-1775
Owner and President: Fred C. Sproul

SANTA CRUZ MOUNTAIN VINEYARD / 2300 Jarvis Road, Santa Cruz, California 95065
(408) 426-6209
Owner and Wine maker: Ken D. Burnap

SCHENLEY DISTILLERS, INC. / 3333 East Church Avenue, Fresno, California 93725
(209) 264-9671
Owner: Schenley Distillers, Inc.,
Vice President and Regional Manager: James A. Steltenpohl

SEGHESIO WINERY, INC. / P. O. Box 24035, Redwood Highway, Cloverdale, California 95425
(707) 857-3581
President: Eugene Seghesio

SHILO VINEYARDS & COMPANY / 8075 Martinelli Road, Forestville, California 95436
(707) 887-2176
Owners: The Shilo Family, President: J. Shilo

SIERRA WINE CORPORATION / Main Office: 555 West Shaw Avenue, Suite B-4, Fresno, California 93704
(209) 227-4067
Owner and President: Berge Kirkorian

SIMI WINERY / Box 946, Healdsburg, California 95448 (707) 433-6981
President: Michael D. Dixon

SONOMA COUNTY CELLARS / P. O. Box 94, Healdsburg, California 95448
(707) 433-1553
Owner: Edith L. Passalacqua

P AND M STAIGER / 1300 Hopkins Gulch Road, Boulder Creek, California 95006
(408) 338-4346
Owners: Paul and Marjorie Staiger

ROBERT STEMMLER WINERY / Lambert Bridge Road, Healdsburg, California 95448
(707) 433-6334

STEVENOT WINERY / San Domingo Road, Murphy, California 95247
(209) 728-3793
Owner and Wine maker: Barden E. Stevenot

STOCKTON DISTILLERY CORP. / P. O. Box 648, Lodi, California 95240
(209) 727-5541
President: Leroy E. Francis

SWAN VINEYARDS / 2916 Laguna Road, Forestville, California 95436
(707) 546-7711
Owner: Joseph A. Swan

ADDENDUM

TEJON MARKETING COMPANY / P. O. BOX 1000, Lebec, California 92343
(805) 327-8481
President: J. E. Morgan

TRADER JOE'S WINERY / 538 Mission Street, South Pasadena, California 91030

UNITED VINTNERS, INC. / 601 4th Street, San Francisco, California 94107
(415) 777-6500
Board Chairman: J. A. Powers

VIE-DEL COMPANY / 11903 S. Chestnut Avenue, P. O. Box 2896, Fresno, California 93745
(209) 834-2525
President: M. S. Nury

VINA VISTA VINEYARD / Winery: Chianti Road, Geyserville, California 95441
Business Office: 2680 Bayshore, Frontage Road, Mountain View, California 94040
(415) 967-1824
President: Keith D. Nelson

VOSE VINEYARDS / 4035 Mt. Veeder Road, Napa, California 94558
(707) 944-2254
Owner and President: Hamilton Vose III

WINE SERVICE COOPERATIVE INC. / 1150 Dowdell Lane, St. Helena, California 94574

WITTWER WINERY / 2440 Frank Avenue, Eureka, California 95501
(707) 443-8852
Owner: J. R. Wittwer

WOODBRIDGE VINEYARD ASSN. / 4614 W. Turner Road, Lodi, California 95240
(209) 369-2614
Owner: a grower cooperative, President: C. L. (Casey) Moore

YERBA BUENA INDUSTRIES / Pier 33, San Francisco, California 94111
(415) 955-9971 Winery: 397-1967
President: Bryan R. R. Whipple

ZACA MESA WINERY / Foxen Canyon Road, Los Olivos, California 93441
(805) 688-3763
President: Louis M. Ream

ZD WINES / Box 900, 20735 Burndale Road, Sonoma, California 95476
(707) 539-9137 or 938-0750
Co-owner: Norman C. de Leuze

Colorado

COLORADO MOUNTAIN VINEYARDS / 15740 W. Sixth Avenue, Golden,
Colorado 80401　　　　　　　　　　　　　　　　　　(303) 278-9463
General Manager: James E. Seewald

Delaware

NORTHMINSTER WINERY / 215 Stone Crop Road, Wilmington,
Delaware 18910　　　　　　　　　　　　　　　　　　(302) 772-2162
Owner: Richard O. Becker

Florida

BARTELS WINERY, INC. / 120 S.I. Street, Pensacola, Florida 32501
(904) 432-8464

FRUIT WINES OF FLORIDA, INC. / 513 South Florida Avenue, Tampa,
Florida 33602　　　　　　　　　　　　　　　　　　　(813) 223-1222
President: Joseph D. Midulla, Sr.

TODHUNTER INTERNATIONAL, INC. / 203–207 Commerce Building, 324
Datura Street, West Palm Beach, Florida 33401　　　(305) 655-8977
President: A. Kenneth Pincourt

Georgia

MONARCH WINE CO. OF GEORGIA / 451 Sawtell Avenue S.E., Atlanta,
Georgia 30315　　　　　　　　　　　　　　　　　　　(404) 622-4496
Owner: Kane-Miller Corp., President: Howard J. Weinstein

Hawaii

TEDESCHI VINEYARD LTD. / P. O. Box 953, Ulupalakua, Maui, Hawaii
96790　　　　　　　　　　　　　　　　　　　　　　　(808) 878-6058
President: Emil P. Tedeschi

Illinois

CONSOLIDATED DISTILLED PRODUCTS, INC. (Union Liquor Co.) / 3247 S. Kedzie, Chicago, Illinois 60623   (312) 254-9000
General Manager: Jerome S. Leavitt

LYNFRED WINERY / 15 South Roselle Road, Roselle, Illinois 60172
(312) 529-1000
Owner and President: Fred E. Koehler

Indiana

HUBER ORCHARD WINERY / Route 1, Box 202, Borden, Indiana 47106
(812) 923-5444
Co-owners: Carl and Gerald Huber

POSSUM TROT VINEYARDS / 8310 N. Possum Trot Road, Unionville, Indiana 47468   (812) 988-2694
Co-owners: Ben and Leora D. Sparks

Iowa

COLONY VILLAGE WINERY / Interstate 80 and Amana, Exit 55, Williamsburg, Iowa 52361   (319) 622-3379
Owners: Les Ackerman and Les Roenig

LITTLE AMANA WINERY, INC. / Interstate 80, 8 miles southwest of Amana, Amana, Iowa 52172

OKOBOJI WINERY INC. / Highway 71, Okoboji, Iowa 51355
Mailing address: Box 449, Arnolds Park, Iowa 51331   (712) 332-2674
President: L. A. Becker, Jr.

OLD STYLE COLONY WINERY, INC. / Third Street, Middle Amana, Iowa 52203   Owner: George Kraus

OLD WINE CELLAR WINERY / Amana, Iowa 52203   (319) 622-3116
Owner and Wine maker: Ramon F. Goerler

PRIVATE STOCK WINERY / 926 Eighth Street, Boone, Iowa 50036
(515) 432-8348
Owners: Tom and Rose Larson

THE WAUKON CORPORATION / 3 Allamakee Street, Waukon, Iowa
52172 (319) 568-3401
President: R. A. Collins

Louisiana

SIG'S ORANGE WINE / P. O. Box 126, Port Sulphur, Louisiana 70083
(504) 564-2333
Owner: Sigvald J. Udstad

Maine

MAIN BOTTLERS (FAIRVIEW WINE CO.) / 701 Water Street, Gardiner,
Maine 04345 (207) 582-4152
President: John J. Jannace

Maryland

BON SPURONZA, INC. / 1522 Stone Road, Westminster, Maryland 21157

ZIEM VINEYARDS / Route 1, Fairplay, Maryland 21733 (301) 223-8352
Owners: Robert and Ruth Ziem

Massachusetts

COMMONWEALTH WINES / Cordage Park, Court Street, Plymouth,
Massachusetts 02360 (617) 746-4138
President: David Tower

Michigan

MILAN WINERIES / 4109 Joe Street, at 6000 Michigan Avenue, Detroit,
Michigan 48210 (313) 894-6464
Board Chairman: I. Murray Jacobs

Mississippi

THOUSAND OAKS VYD. WINERY / Route 4, Box 133, Starkville, Mississippi 39701 (601) 323-6657
Owner and General Manager: Robert M. Burgin

THE WINERY RUSHING / Merigold, Mississippi 38759 (601) 748-3821
Co-owner: O. W. Rushing

Missouri

AXEL N. ARNESON & N. A. ARNESON (PEACEFUL BEND VINEYARD) / Route 2, Steelville, Missouri 65565

ASHBY VINEYARDS, INC. (ROSATI WINERY) / Route 1, Box 55, St. James, Missouri 65559 (314) 265-8629
Owner: Ashby Vineyards, Inc., President: Robert Ashby

GARCO WINE COMPANY / 4017-31 Folsom Avenue, St. Louis, Missouri 63110 (314) 664-8300
Owner: Jack Cohen

HERMANNHOF / 330 E. First Street, Hermann, Missouri 65041
President: James Dierberg

MIDI VINEYARDS, LTD. / Route 1, Lone Jack, Missouri 64070
Office: 1339 E. 109 Terrace, Kansas City, Missouri 64131
Co-owner: Dutton Biggs

New Jersey

B & B VINEYARDS INC. / R.D. #2, Box 235, Stockton, New Jersey 08559 (201) 996-6353
President: John A. Boyd

JACOB LEE WINERY / Route 130, R.F.D. 1, Bordentown, New Jersey 08505 (609) 298-4860
Owner: J. Jacob Lee

POLITO VINEYARD / Route 4, Box 630, Vincentown, New Jersey 08088
(609) 268-0537
Owners: Frank and Dorothy Polito

JOHN SCHUSTER & SON / 502 St. Louis Avenue, Egg Harbor City, New Jersey 08215   (609) 965-0223
Owner: John J. Schuster

New Mexico

CORRALES BONDED WINERY / Box 302B, Corrales, New Mexico 87048
(505) 898-2904
Owner: Tony R. Gonzales

ESTRADA WINERY / P. O. Box 202, Mesilla, New Mexico 88046
Owner: Dan M. Estrada

MIGUEL O. GONZALES / Box 22, Dona Ana, New Mexico 88032

LA VINA / Box 121, Chamberino, New Mexico 88027   (505) 882-2092
President: Clarence Cooper

RICO'S WINERY / 6406 N. 4th Street, N.W., Albuquerque, New Mexico 87107   (505) 344-2075
Owner: Enrico Gradi

New York

CANANDAIGUA WINE COMPANY, INC. / 116 Buffalo Street, Canandaigua, New York 14424   (716) 394-3630
President: Marvin Sands

DE MAY WINE CELLARS / Hammondsport, New York 14840
(607) 569-2040
Owners: Serge and Monique De May

DISTILLERIE STOCK U.S.A., LTD. / 58-58 Laurel Hill Boulevard, Woodside, New York 11377   (212) 651-9800
President: Mario Morpurgo

EL PASO WINERY / R.D. 1, Box 170, Ulster Park, New York 12487

FREDONIA PRODUCTS CO., INC. / Cliffstar Avenue, Dunkirk, New York 14048   (716) 366-6100
President: Herbert Star

GALANTE'S FARM WINERY / 9813 Erie Road, Angola, New York 14006
(716) 549-0634

ADDENDUM

GREAT RIVER WINERY & MARLBORO CHAMPAGNE CELLARS / 104 Western Avenue, Marlboro, New York 12542 (914) 236-2521

HAMMONDSPORT WINE CO., INC. / 89 Lake Street, Hammondsport, New York 14840 (607) 569-2255
President: George H. Page

LONG ISLAND VINEYARDS, INC. / Alvah's Lane, P. O. Box 927, Cutchogue, New York 11935 (516) 734-5158
President: Alexander Hargrave

LOUKAS WINES / 910 Gerard Avenue, Bronx, New York 11452

MACK FRUIT FARMS / 1593 Hamlin Parma Town Line Road, Hilton, New York 14468

NORTHLAKE VINEYARDS / R.D. 1, Box 271, Romulus, New York 14541 (607) 273-6804
Owners: Fred and Marilyn Williams

OLD DUTCH MUSTARD CO., INC. / 80 Metropolitan Avenue, Brooklyn, New York 11211 (212) 387-9155

VALLEY VINEYARDS / Oregon Trail Road, Walker Valley, New York 12588 (914) 744-3912
Owner: Gerrit Dross

WAGNER VINEYARDS / Lodi, New York 14860 (607) 582-6450
Owner: Stanley A. Wagner

WINDMILL FARMS / 193 County Line Road, Ontario, New York 14519

North Carolina

THE BILTMORE COMPANY / Biltmore Estate, Asheville, North Carolina 28803

JOHN C. DOCKERY, JR. / Cheraw Highway, Rockingham, North Carolina 28379 (919) 895-4934

DUPLIN WINE CELLARS, INC. / Rose Hill, North Carolina 23458
(919) 289-3888
Owner: a cooperative, President: Dan Fussell

Ohio

AMERICAN VINEYARDS CO., INC. / 2220 Center Street, Cleveland, Ohio
44113 (216) 241-4190
President: William F. Hauck

LESLIE J. BRETZ / P. O. Box 17, Middle Bass, Ohio 43446
(419) 285-2323
Owner: Leslie J. Bretz

BRUSHCREEK VINEYARDS / R.R. 4, 12351 Newkirk Lane, Pebbles,
Ohio 45660 (513) 588-2618
Owners: Ralph A. and Laura J. Wise

BUCCIA VINEYARDS / 518 Gore Road, Conneaut, Ohio 44030
(216) 593-5976
Owner: Alfred J. Bucci

CATAWBA ISLAND WINE CO. (Mon Ami Champagne Co.) / 326 West
Catawba Road, Catawba Island, Port Clinton, Ohio 43452 (419) 797-4445
President: Norman E. Mantey

JOHN CHRIST WINERY / 32421 Walker Road, Avon Lake, Ohio 44012
(216) 933-3046

COHODAS VINEYARDS, INC. / R.F.D. 2, County Line Road, Geneva,
Ohio 44041 (216) 466-3574
Owner and President: Morris A. Cohodas

COLONIAL VINEYARDS / 6222 N. State Route 48, Lebanon, Ohio 45036
(513) 932-3842
Owner: Norman E. Greene

HERITAGE VINEYARDS / 6020 South Wheelock Road, West Milton,
Ohio 45383 (513) 698-5369
Owner: Edward W. Stefanko

EDWARD W. JOHLIN / 3935 Corduroy Road, Oregon, Ohio 43616
(419) 693-6288

LE BOUDIN VINEYARD & VINERY / R.R. #3, Box 172, Cardington, Ohio
43315 (419) 768-2091
Owner: Ruth W. Hubbell

CARL M. LIMPERT / 28083 Detroit Road, Westlake, Ohio 44091
(216) 871-0035
Owner: Carl M. Limpert

LUKENS VINEYARD, INC. / 2446 School Road, Hamilton, Ohio 45103

McINTOSH'S OHIO VALLEY WINES / R.R. 1, Box 190, Bethel, Ohio
45106 (513) 379-1159
Officer: Charles D. McIntosh

POMPEI WINERY, INC. / 3995 East 86th Street, Cleveland, Ohio 44105
President: Frederick De Pompei

RINI WINE COMPANY / 1854 Scranton Raod, Cleveland, Ohio 44113
(216) 771-5733
President: Edmund Thomas

SHAWNEE VINEYARDS / R.R. 1, Circleville, Ohio 43113
Proprietor: Jack Goode

TARULA FARMS / 1786 Creek Road, Clarksville, Ohio 45113
(513) 289-2181
Executive Assistant: Vickie McClary

WICKLIFFE WINERY / 29555 (Rear) Euclid Avenue, Wickliffe, Ohio
44092

Oklahoma

RHEINFRANK CELLARS / Route 1, Box 179-A, Cleveland, Oklahoma
74020 Owner: Dr. Robert E. Rheinfrank

PETE SCHWARZ WINERY / Box 545, Okarche, Oklahoma 73762
(405) 263-7664
Owner: Pete Schwarz

Oregon

BIG FIR WINERY / 1612 W. Division Street, Gresham, Oregon 97030

CENTURY HOME WINE / Route 2, Box 111, Newberg, Oregon 97132
(503) 538-9743
Owner: Nellie M. Maze

COTE DES COLOMBE VINEYARDS / P. O. Box 266, Banks, Oregon
97106 (503) 646-1223
Owner: Cote des Colombe Vineyard, Inc.,
President: Joseph R. P. Coulombe

HENRY ENDRES WINERY / 13300 S. Clackamas River Drive, Oregon
City, Oregon 97045 (503) 656-7239
Owner: Henry C. Endres

HENRY'S WINERY / Box 26, Umpqua, Oregon 97486 (503) 459-5120
Owner and President: Scott Henry

THE HONEY HOUSE WINERY / 26202 Fawver Road, Veneta, Oregon
97487 (503) 935-2008
Owner: Robert H. Saxton

HONEYWOOD FRUIT WINERY / P. O. Box 12278, Salem, Oregon 97309
(503) 362-4111
Owner: Northwest Distillers, President: Paul Gallick, Jr.

HUMBUG WINERY / 300 Hidden Valley Lane, Roseburg, Oregon 97470
Owners: Dr. and Mrs. T. R. Mafit

MT. HOOD WINERY / P. O. Box 65, Mt. Hood, Oregon 97041
(503) 352-6465
General Manager: Lester J. Martin

MT. TABOR WINE CO. / 7234 Southeast Main Street, Portland, Oregon
97215

REUTER'S HILL VINEYARDS / P. O. Box 883, Forest Grove, Oregon
97116 (503) 357-5513
President: William Miller

SISKIYOU VINEYARDS / 6220 Caves Highway, Cave Junction, Oregon
97253 (503) 592-3727
Co-owner: Charles W. David

VALLEY VIEW VINEYARDS / 1352 Applegate Road, Jacksonville,
Oregon 97530 (503) 899-8468
Owners: Frank and Ann Wisnovsky

Pennsylvania

BUFFALO VALLEY WINERY / R.D. 2, Buffalo Road, Lewisburg,
Pennsylvania 17837 (717) 524-2143
Owner and President: Charles W. Pursel

DOERFLINGER WINE CELLARS, INC. / 3248 Old Berwick Road,
Bloomsburg, Pennsylvania 17815 (717) 784-2112
President: Ludwig A. Doerflinger

ADDENDUM

HERITAGE WINE CELLARS / 12162 E. Main Road, North East,
Pennsylvania 16428 (814) 725-8015
Owners: Robert C., Michael N., and William M. Bostwick

MERLINO'S WINERY / 22 Myrtle Avenue, Springfield, Pennsylvania
19070 (215) 544-6457

MONSEY CORPORATION T/A PEQUEA VALLEY VINEYARD &
WINERY / Box 332, Willow Street, Pennsylvania 17584 (717) 464-3721
President: H. Peterman Wood

NAYLOR WINE CELLARS, INC. / R.D. 3, Stewartstown, Pennsylvania 17363
Mailing address: R.D. 2, Box 85, York, Pennsylvania 17403 (717) 741-1236
Owners: Richard and Audrey Naylor

NISSLEY VINEYARDS / Route #1, Bainbridge, Pennsylvania 17502
(717) 426-3514
President: J. Richard Nissley

Rhode Island

SOUTH COUNTY VINEYARDS, INC. / Brow Lane (P. O. Box 2), Slocum,
Rhode Island 02877 (401) 294-3100
Owners: South County Vineyards, Inc., President: H. Winfield Tucker

South Carolina

TENNER BROTHERS, INC. / R.F.D. 2, Box 85, Patrick, South Carolina
29584 (803) 634-6621
President: Marvin Sands

Texas

GUADALUPE VALLEY WINERY / 1720 Hunter Road, New Braunfels,
Texas 78130 (512) 629-2351
President: William J. Gallagher

LA BUENA VIDA VINEYARDS / WSR Box 18-3, Springtown, Texas 76082
(817) 523-4366
Owner: Bobby G. Smith

LLANO ESTACADO WINERY / P. O. Box 6170, Lubbock, Texas 79413
Winery: 3.2 miles east of U.S. 87 on FM 1585   (806) 745-2258
President: C. M. McPherson

VAL VERDE WINERY / 139 Hudson Drive, Del Rio, Texas 78840
(512) 775-9714
Owner: Thomas M. Qualia

Virginia

LAIRD & COMPANY / North Garden, Virginia 22959   (804) 296-6058
Owner: Laird & Co., Board Chairman: J. E. Laird

PIEDMONT WINERY / P. O. Box 286, Middleburg, Virginia 22117
(703) 687-5134
Owner and President: Mrs. Thomas Furness

RICHARD'S WINE CELLARS, INC. (See Canandaigua Wine Co.,
Inc.) / 120 Pocahontas Street, Petersburg, Virginia 23803   (804) 733-6786
President: Marvin Sands

SHENANDOAH VINEYARDS / Route 2, Box 208 B, Edinburg, Virginia
22824   (703) 984-8699
Owner: J. B. Randel

THE VINEYARD / State Route 739, Winchester, Virginia 22601
Mailing address: Route 5, Box 486 Y, Winchester, Virginia 22601
Owner: Robert Viehman

WOBURN WINERY / R.F.D. 1, Clarksville, Virginia 23927
Owner: John and June Lewis

Washington

ALHAMBRA WINE COMPANY / P. O. Box 338, Selah, Washington 98942
(509) 697-7292
Owner: Otis B. Harlan

BINGEN WINE CELLARS, INC. / 315 W. Steuben, Bingen, Washington
98605   (509) 493-3001
President: Charles V. Henderson

ADDENDUM

PRESTON WINE CELLARS / Star Route 1, Box 1234, Pasco, Washington 99301
(509) 545-1990
Owner: S. W. "Bill" Preston

SNOHOMISH VALLEY WINERY, INC. / 1211 2nd Street, Marysville, Washington 98270
(206) 659-9858
President: Dr. Les Venables

Wisconsin

DOOR-PENINSULA WINE INC. / Route 1, Sturgeon Bay, Wisconsin 54235
(414) 743-7431
President: Evangeline Alberts

FRESNO WINE CO. OF MILWAUKEE, INC. / 448 East Bruce Street, Milwaukee, Wisconsin 53204
(414) 271-5524
President: Kathleen Pasamani

FRUIT OF THE WOODS WINE CELLAR, INC. / 1113 Wall Street, Eagle River, Wisconsin 54521
(715) 479-4800
President: John McCain

VON STIEHL WINE, INC. / P. O. Box 45, 115 Navarino Street, Algoma, Wisconsin 54201
(414) 487-5208
President: C. W. Stiehl

Index of Vineyards

A. Nonini Winery, 176
A. Rafanelli, 83
Ackerman Winery, 162
Adams County Winery, 147
Ahern Winery, 170
Ahlgren Vineyards, 20
Alatera Vineyards, 170
Alexander Valley Vineyards, 70
Alexis Bailly Vineyard, Inc., 108–109
Alhambra Wine Company, 190
Almadén Vineyards, 20–21
Amador Winery, 170
American Vineyards Company, Inc., 186
Amity Vineyards, 90
Andre's Wines, Ltd., 124
Andre's Wines (B.C.) Ltd., 97
Antuzzi's Winery, 150
Argonaut, 40
Arneson, N. A., see Axel N. Arneson & N. A. Arneson (Peaceful Bend Vineyards)
Associated Vintners, Inc., 94–95
Au Provence Restaurant and Cedar Hill Wine Company, 118
Axel N. Arneson & N. A. Arneson (Peaceful Bend Vineyards), 166–67, 183

B. Cribari & Sons Winery, 24
Balic Winery, 150
Ballard Canyon Corp. Winery, 170
Bandiera Wines, 170
Banholzer Winecellars, Ltd., 117
Barboursville Corporation, 156
Bardenheier Wine Cellars, 164
Barengo Vineyards, 21
Bargetto's Santa Cruz Winery, 21
Barry Wine Company, The, Inc. (O-Neh-Da Vineyards), 128–29
Bartels Winery, Inc., 180
B & B Vineyards Inc., 183
Beaulieu Vineyard, 54, 170
Beckett Cellars (John), 54
Beliz-De Loach Vineyards, 170
Bella Napoli Winery, 170
Benmarl Wine Company, 140
Beringer Vineyards, 55
Bernardo Winery, 14
Berrywine Plantation, 152
Bertero Winery, The, 22
Big Fir Winery, 187
Biltmore Company, The, 185
Bingen Wine Cellars, Inc., 190
Bjelland Vineyards, 90
Bodego de Rancho Viejo, S.A., 17–18
Bodegas de Santo Tomas, S.A., 18
Boeger Winery, 40–41
Bon Spuronza, Inc., 182
Boordy Vineyards, 152
Borra's Cellar, 170
Boskydel Vineyard, 110

Bowman Wine Cellars, 164
Bretz, Leslie J., 186
Bronco Wine Company and JFJ Winery, 41
Bronte Winery and Vineyards, 111
Brookside Enterprises, Inc., 170
Brotherhood Winery, The, 141
Bruce Winery (David), 22
Brushcreek Vineyards, 186
B & R Vineyards, Inc., 170
Buccia Vineyards, 186
Buckingham Valley Vineyards, 147
Bucks County Vineyards, 147-48
Buehler Vineyards, 171
Buena Vista Winery, 71
Buffalo Valley Winery, 188
Bully Hill Vineyards, 129
Burgess Cellars, 55
Bynum Winery (Davis), 171
Byrd Vineyards, 152-53

Cadenasso Winery, 171
Cadlolo Winery, 41
Cagnasso Winery, 141
Cakebread Cellars, 55
Calera, 22
California Cellar Masters, 41
California Concentrate Company, 23
California Growers Winery Inc., 171
California Wine Association, 23
Callaway Vineyard and Winery, 14-15
Calona Wines Ltd., 97-98
Cambiaso Vineyards, 71
Canandaigua Wine Company, Inc., 184, 190
Cannery Wine Cellar, The, 171

Carey Winery (Richard), 42
Carl M. Limpert, 186
Carmel Bay Winery, 171
Carneros Creek Winery, 55-56
Casabello Wines, Ltd., 98
Casa de Fruta, 23
Cascade Mountain Vineyards, 141
Cassayre-Forni Cellars, 56
Catawba Island Wine Company (Mon Ami Champagne Company), 186
Caymus Vineyards, 56
Cedar Hill Wine Company, 118
Center Ridge Winery, 169
Century Home Wine, 187
Chalet Debonné Vineyards, Inc., 118-19
Chalone Vineyard, 24
Channing Rudd Cellars, 177
Chappellet Winery, 171
Charal Winery & Vineyards, Inc., 125
Charles Coury Vineyards, 91
Charles Krug Winery, 56, 174
Château Chevalier Winery, 56
Château Grand Travers, Ltd., 111
Château Montelena Winery, 171
Château Piatt Winery, Inc., 148
Château St. Jean, 72
Château Ste. Michelle, 95
Château Sonoma Winery, 72
Cherpin Winery (Louis), 171
Chicama Vineyards, 135
Chispa Cellars, 171
Christian Brothers, 57
Christina Wine Cellars, 162
Christ Winery (John), 186
Clinton Vineyards, 142
Clos du Bois, 72-73
Clos du Val, 57
Coarsegold Wine Cellar, 171

INDEX OF VINEYARDS

Cohodas Vineyards, Inc., 186
Colcord Winery, 106
Colonial Vineyards, 186
Colony Village Winery, 181
Colony Wine and Cheese Haus, 162
Colorado Mountain Vineyards, 180
Commonwealth Wines, 182
Concannon Vineyard, 42
Conestoga Vineyards, 148
Congress Springs Vineyards, 24
Conn Creek Winery, 57
Conrad Viano Winery, 48–49
Conrotto Winery, 171
Consolidated Distilled Products, Inc., 181
Consolidated Distribution Services, Inc., 171
Consumnes River Vineyard, 172
Corrales Bonded Winery, 184
Cote des Colombe Vineyards, 187
Coury Vineyards (Charles), 91
Cowie Wine Cellars, 169
Cresta Blanca Winery, 73
Cribari & Sons Winery (B.), 24
Cucamonga Vineyard Company, 172
Cucamonga Vintners, 172
Cuvaison, 57–58
Cygnet Cellars, 24–25

Dach Ranch (Fruit Stand), 73
Dach Vineyards, 73
D'Agostini Winery, 172
David Bruce Winery, 22
Davis Bynum Winery, 171
Dehlinger Winery, 172
Delano Growers Co-operative Winery, 172

Delicato Vineyards, 42
Del Rey Winery, 172
De May Wine Cellars, 184
De Salvo's Winery, 169
Diablo Vista, 43
Diamond Creek Vineyards, 58
Digardi Winery, 172
Distillerie Stock U.S.A., Ltd., 184
Dockery, John C., Jr., 185
Doerflinger Wine Cellars, Inc., 188
Domaine Chandon, 58
Door-Peninsula Wine Inc., 191
Dover Vineyards, 119
Dry Creek Vineyard, 73–74
Duplin Wine Cellars, Inc., 185
Durney Vineyard, 172
Dutch Country Wine Cellar, 148–49

Easley Winery, 104
East-Side Winery, 43
Edmeades Vineyards, 74
Edward W. Johlin, 186
Ehrle Brothers, 163
E and K Wine Company, 119
Eldorado Vineyards, 43
Elk Cove Vineyards, 91
El Paso Winery, 184
Emilio Guglielmo Winery, 27
Endres Winery (Henry), 188
Enz Vineyards, 172
Erbacher Vineyards, 134
Estrada Winery, 184
Estrella River Winery, 172
Eyrie Vineyards, 91

F. J. Miller Winery, 62
F. Korbel and Brothers, 79

Fairview Wine Company, *see* Main Bottlers (Fairview Wine Company)
Farfelu Vineyard, 157
Farnesi Winery, 172
Felton-Empire Vineyards, 25
Fenn Valley Vineyards and Wine Cellar, 111–12
Ferrara Winery, 15
Ferreira Wines, Inc., 173
Fetzer Vineyards, 74
Ficklin Vineyards and Winery, 25
Fieldbrook Valley Winery, 74–75
Field Stone Winery, 75
Fink Winery, 112
Firestone Vineyard, 25
Firpo Winery, 173
Foppiano Vineyards, 75
Forgeron Vineyards, 91–92
Fortino Winery, 26
Franciscan Vineyards, 59
Franzia Brothers, 26
Frasinetti and Sons (James), 173
Fredonia Products Company, Inc., 184
Fredson Winery, 173
Freemark Abbey, 59
Fresno Wine Company of Milwaukee, Inc., 191
Fretter Wine Cellars, 173
Freyaldenhoven's Winery, 169
Frick Winery, The, 26
Frontenac Vineyards, Inc., 112
Fruit Wines of Florida, Inc., 180
Fruit of the Woods Wine Cellar, Inc., 191

Galante's Farm Winery, 184
Galleano Winery, 15–16
Gallo Vineyards, 43–44

Garco Wine Company, 183
Giretti, Peter and Harry, 173
Gauthier & Clevenger Ltd., 173
Gem City Vineland Company, Inc., 115
Gemello Winery, The, 44
Geyser Peak Winery, 75–76
Gibson Wine Company, 44
Giumarra Vineyards, 27
Giuseppe Mazzoni, 175
Glenora Wine Cellars, 130
Golden Rain Tree Winery, 104
Gold Seal Vineyards, Inc., 130
Gonzales, Miguel O., 184
Grand Cru Vineyards, 76
Grand Pacific Vineyard Company, 76
Grand River Vineyard, 119–20
Great River Winery, 142
Great River Winery & Marlboro Champagne Cellars, 185
Great Western, *see* Pleasant Valley Wine Company
Green & Red Vineyard, 173
Green Valley Vineyards, 164–65
Grgich Hills Cellar, 173
Gross Highland Winery, 151
Guadalupe Valley Winery, 189
Guglielmo Winery (Emilio), 27
Guild Wineries and Distilleries, 173
Guilford Farm Vineyard, 157
Gundlach-Bundschu Winery, 76–77

Hacienda Wine Cellars, 77
Hafle Vineyards, 101
Hamlet Hill Vineyard, 134–35
Hammondsport Wine Company, Inc., 185
Hanns Kornell Champagne Cellars, 60–61, 174

INDEX OF VINEYARDS

Hanzell Vineyards, 77
Haraszthy & Son (J. J.), 173
Harbor Winery, 44, 173
Hargrave Vineyard, 142
Hecker Pass Winery, 27
Heckmann's Winery, 169
Heineman Winery, 120
Heitz Wine Cellars, 59-60
Henry Endres Winery, 188
Henry J. Sax, 169
Henry's Winery, 188
Heritage Vineyards, 186
Heritage Wine Cellars, 189
Hermannhof, 183
Heron Hill Vineyards, 130
Heublein, Inc., 173
Hillcrest Vineyard, 92
Hinzerling Vineyards, 95
Hoffman Mountain Ranch Vineyard, 27-28
Honey House Winery, The, 92, 188
Honeywood Fruit Winery, 188
Hop Kiln Winery at Griffin Vineyards, 77-78
Horizon Winery, 174
Huber Orchard Winery, 181
Hudson Valley Wine Company, Inc., 143
Humbug Winery, 188
Husch Vineyards, 78

Inglenook Vineyards, 60
Italian Swiss Colony, 78

J. J. Haraszthy & Son, 173
J. Mathews Napa Valley Winery, 61-62
J. Pedroncelli Winery, 82
J. W. Morris Port Works, 45

Jacob Lee Winery, 183
Jade Mountain Winery, 174
James Frasinetti and Sons, 173
Jekel Vineyards, 174
Johlin, Edward W., 186
John Beckett Cellars, 54
John C. Dockery, Jr., 185
John Christ Winery, 186
John Schuster & Son, 184
Johnson Estate Winery, 123
Johnson's Alexander Valley, 78-79
Jonicole Vineyards, 92
Jordan Wines, 125
Joseph Phelps Vineyards, 63

Kedem Royal Winery, 143
Keenan Winery (Robert), 60
Kenwood Vineyards, 79
Kirigin Cellars, 28
Klingshirn Winery, 120
Knudson Erath Winery, 93
Konocti Cellars, 174
Konstantin D. Frank and Sons Vinifera Wine Cellars, Ltd., 129
Korbel and Brothers (F.), 79
Kornell Champagne Cellars (Hanns), 60-61
Kruger's Winery and Vineyards, 165
Krug Winery (Charles), 56
Kruse Winery (Thomas), 28

L. Mawby Vineyards and Winery, 113
La Abra Farm and Winery, 157
La Buena Vida Vineyards, 189
Laird & Company, 190
Lakeside Vineyard, Inc., 112-13

Lake Sylvia Vineyard, 109
Lambert Bridge, 79–80
Lamb Winery and Vineyards (Ronald), 28–29
La Mont Winery, Inc., 174
Landis Vineyard, 174
Landmark Vineyards, 80
Lapic Winery, 122
La Purisima Winery, 174
Las Tablas Winery, 37
La Vina, 184
Le Boudin Vineyard and Vinery, 186
Leelanau Winecellars, Ltd., 113
Lembo Vineyards, 149
Leonetti Cellars, 95
Leslie J. Bretz, 186
Liberty Winery, Inc., 174
Limpert, Carl M., 186
Little Amana Winery, Inc., 181
Live Oaks Winery, 29, 174
Llano Estacado Winery, 190
Llords & Elwood Winery, 29
Lodi Vintners, Ltd., 174
Long Island Vineyards, Inc., 185
Long Vineyards, 174
Los Alamos Winery, 175
Louis Cherpin Winery, 171
Louis M. Martini Winery, 61
Loukas Wines, 185
Lower Lake Winery, 61
Lucas Home Wine/D. Lucas Vineyard, 175
Lukens Vineyard, Inc., 186
Lynfred Winery, 181
Lytton Springs, 80

McIntosh's Ohio Valley Wines, 187

Mack Fruit Farms, 185
Main Bottlers (Fairview Wine Company), 182
Manfred J. Vierthaler Winery, 96
Mantey Vineyards, Inc., 120–21
Marine Wholesale & Warehouse Company, 175
Markham Winery, 175
Markko Vineyard, 121
Mark West Vineyards and Winery, 80
Marlboro Champagne Cellars, 144
Martinelli & Company (S.), 175
Martini and Prati Wines, Inc., 81
Martini Winery (Louis M.), 61
Martin Ray Winery, 33
Martin Winery, Inc., 16
Masson Vineyards (Paul), 29–30
Mastantuono Vineyard, 30
Matanzas Creek Winery, 175
Mathews Napa Valley Winery (J.), 61–62
Mawby Vineyards and Winery (L.), 113
Mayacamas Vineyards, 62
Mazza Vineyards, 122
Meier's Wine Cellars, Inc., 102
Meredyth Vineyards (at Stirling Farm), 157–58
Merlino's Winery, 189
Merritt Estate Winery, 123–24
Michael T. Parsons Winery, 32
Midi Vineyards, Ltd., 183
Miguel O. Gonzales, 184
Milano Winery, 81
Milan Wineries, 182
Mill Creek Vineyards and Winery, 81
Miller Winery (F. J.), 62
Mirassou Vineyards, 30

INDEX OF VINEYARDS

Missouri State Fruit Experiment Station, 165
Mogen David Wine Corporation, 116, 124
Mon Ami Champagne Company, 186
Monarch Wine Company of Georgia, 180
Mondavi Winery (Robert), 62
Monsey Corporation T/A Pequea Valley Vineyard and Winery, 149, 189
Montbray Wine Cellars, Ltd., 153
Montclair Winery, 44-45
Montelle Vineyards, 165-66
Monterey Peninsula Winery, 30-31
Monterey Vineyard, 31
Montevina Winery, 45
Mont La Salle Vineyards, 175
Morris Port Works (J. W.), 45
Mount Bethel Winery, 169
Mount Eden Vineyards, 45
Mt. Hood Winery, 188
Mount Palomar Winery, 16
Mount Pleasant Vineyards, 166
Mount Renaissance Vineyard & Winery, 175
Mt. Tabor Wine Company, 188
Mt. Veeder Winery and Vineyards, 175
Moyer Vineyards, 102

Napa Valley Cooperative Winery, 175
Napa Vintners, 175
Napa Wine Cellars, 175
Nauvoo State Historic Site and Park Vineyard, 116
Navarro Vineyards, 81
Naylor Wine Cellars, Inc., 189
Nehalem Bay Winery, 93
Neil's Winery, Inc., 169
Nicasio Vineyards, 31
Nichelini Vineyard, 63
Nicholas G. Verry, Inc., 37
Niebaum-Coppola Estates, 176
Nissley Vineyards, 149, 189
Noble Vineyards, Inc., 176
Nonini Winery (A.), 176
Nordman of California, 176
Northeast Vineyard, 144
Northlake Vineyards, 185
Northminster Winery, 180
Novitiate Wines, 45-46

Oak Barrel Winery, 46
Oak Glen Winery, 176
Oak Knoll Winery, 93
Oakview Plantations, 159
Obester Winery, 31
Ocean Shores Winery, 96
Okoboji Winery Inc., 181
Old Dutch Mustard Company, Inc., 185
Old Style Colony Winery, Inc., 181
Old Wine Cellar Winery, 181
Oliver Winery, 105
O-Neh-Da Vineyards, *see* Barry Wine Company, The
Opici Winery, Inc., 176
Ozark Vineyards, 166

Page Mill Winery, 46
Papagni Vineyards, 32
Parducci Wine Cellars, 82
Parsons Winery (Michael T.), 32

Pastori Winery, 176
Patowmack Vineyard, 158
Paul Masson Vineyards, 29–30
Peaceful Bend Vineyards, 166–67, 183
Pecota Winery (Robert), 176
Pedrizzetti Winery, 32
Pedroncelli Winery (J.), 82
Pellegrini Bros. Wines, Inc., 176
Penn-Shore Vineyards, Inc., 122
Pequea Valley Vineyard & Winery, see Monsey Corporation T/A Pequea Valley Vineyard & Winery
Perelli-Minetti & Sons, 176
Pesenti Vineyard, 32
Peter and Harry Giretti, 173
Pete Schwarz Winery, 187
Phelps Vineyards (Joseph), 63
Piedmont Winery, 190
Pirrone Wine Cellars, 176
Pleasant Valley Wine Company (Great Western), 131
P and M Staiger, 178
Pocai & Sons, 176
Polito Vineyard, 183
Pompei Winery, Inc., 187
Ponzi Vineyards, 93–94
Pope Valley Winery, 82
Possum Trot Vineyard, 105, 181
Post Winery Inc., 169
Presque Isle Wine Cellars, 123
Prestige Vineyards, 176
Preston Wine Cellars, 191
Preston Winery, 83
Private Stock Winery, 181
Provenza Vineyards, 153
Prudence Island Vineyards, 135–36
Puyallup Valley Winery, 96

Quady Winery, 33

Rafanelli (A.), 83
Rancho de Philo, 16
Rancho Sisquoc Winery, 33
Rauner and Sons Winecellars and Vineyard, 117
Raymond Vineyard and Cellar, 63
Ray Winery (Martin), 33
Rege Wine Company, 177
Renault Winery, 151
Reuter's Hill Vineyards, 188
Rheinfrank Cellars, 187
Richard Carey Winery, 42
Richard Minor Winery, see Ritchie Creek Vineyards
Richard's Wine Cellars, Inc., see Canandaigua Wine Company, Inc.
Richert & Sons Winery, 33
Rico's Winery, 184
Ridge Vineyards, 34
Rini Wine Company, 187
Rio Vista Winery, Inc., 177
Ritchie Creek Vineyards, 177
River Run Vintners, 177
RME, Inc., 177
Robert Keenan Winery, 60
Robert Mondavi Winery, 62
Robert Pecota Winery, 176
Robert Stemmler Winery, 178
Robin Fils & Cie, Ltd., 131
Ronald Lamb Winery and Vineyards, 28–29
Rosati Winery, 167
Rosenblum Cellars, 177
Roudon Smith Vineyards, 35
Round Hill Cellars, 177
Rubidoux Winery, 177

INDEX OF VINEYARDS

Rutherford Hill Winery, 63–64
Rutherford Vintners, 64

S. Martinelli & Company, 175
Ste. Chapelle Vineyards, 97
St. Clement Vineyards, 66
St. Helena Wine Company, 177
St. James Winery, 167
St. Julian Wine Company, Inc., 113–14
Ste. Michelle Wines, 98
Sakonnet Vineyards, 136
San Antonio Winery, Inc., 177
San Benito Vineyards, Inc., 177
Sandstone Winery, 163
Sanford & Benedict Vineyard, 34
San Joaquin Warehousing Company, 178
San Martin Winery, 34
San Pasqual Vineyards, 17
Santa Barbara Winery, 34–35
Santa Cruz Mountain Vineyard, 178
Santa Ynez Valley Winery, 35
Sattui Winery (V.), 64
Sausal Winery, 83
Sax, Henry J., 169
Schapiro Wine Company, 144
Schenley Distillers, Inc., 178
Schiehallion Vineyards, 158
Schramsberg Vineyards, 64
Schuster & Son (John), 184
Schwarz Winery (Pete), 187
Sebastiani Vineyards, 83–84
Seghesio Winery, Inc., 178
Sequoia Cellars, 46–47
Shawnee Vineyards, 187
Shenandoah Vineyards, 47, 190
Sherrill Cellars, 35

Shilo Vineyards & Company, 178
Sierra Vista, 47
Sierra Wine Corporation, 178
Sig's Orange Wine, 182
Silver Oak Cellars, 65
Simi Winery, 84, 178
Siskiyou Vineyards, 188
Smith-Madrone Vineyards, 65
Smothers, 35–36
Snohomish Valley Winery, Inc., 191
Sokol Blosser Winery, 94
Sommelier, 36
Sonoma County Cellars, 178
Sonoma County Cooperative Winery, 84
Sonoma Vineyards, 84–85
Sotoyome Winery, 85
South Coast Cellars, 17
South County Vineyards, Inc., 189
Souverain Cellars, 85
Spring Mountain Vineyards, 65
Stag's Leap Wine Cellars, 65
Stag's Leap Winery, 66
Stemmler Winery (Robert), 178
Sterling Vineyards, 66
Steuk Wine Company, 121
Stevenot Winery, 178
Stirling Farm, see Meredyth Vineyards (at Stirling Farm)
Stockton Distillery Corporation, 178
Stoltz Vineyards and Winery, 167
Stonegate Winery, 66–67
Stone Hill Wine Company, 168
Stone Mill Winery, Inc., 109
Stoneridge, 47
Stony Hill Vineyard, 67
Stony Ridge Winery, 47–48
Sublette Winery (Warren J.), 102–103

Sunrise Winery, 36
Sutter Home Winery, 67
Swan Vineyards, 178
Swiss Valley Vineyard, 105
Sycamore Creek Vineyards, 36

Tabor Hill Vineyard and
  Winecellar, Inc., 114
Tarula Farms, 187
Taylor Wine Company, 131
Tedeschi Vineyard Ltd., 180
Tejon Marketing Company, 179
Tenner Brothers, Inc., 189
Thomas Kruse Winery, 28
Thomas Vineyards, 17
Thompson Winery, 117
Thousand Oaks VYD. Winery, 183
Todhunter International, Inc., 180
Tomasello Winery, 151
Trader Joe's Winery, 179
Trefethen Vineyards, 67–68
Trentadue Winery, 85–86
Truluck Winery, 160
Tualatin Vineyards, Inc., 94
Tulocay Winery, 68
Turgeon and Lohr Winery, 48

United Vintners, Inc., 179

V. Sattui Winery, 64
Valley of the Moon Vineyards, 86
Valley View Vineyards, 188
Valley Vineyards, 185
Valley Vineyards Farm, Inc., 103
Val Verde Winery, 190
Veedercrest Winery, 48
Vega Vineyards, 37

Vendramino Vineyards Company,
  114–15
Ventana Vineyards and Winery,
  37
Verry, Nicholas G., 37
Viano Winery (Conrad), 48–49
Vides de Guadalupe (Domeq), 18
Vie-Del Company, 179
Vierthaler Winery (Manfred J.), 96
Villa Armando, 49
Villa Bianchi Winery, 38
Villa D'Ingianni Winery, 132
Village Winery, 163
Villa Medeo Vineyards, 118
Villa Mount Eden, 68
Vina Vista Vineyard, 179
Vineyard, The, 190
Vinifera Wine Cellars, *see*
  Konstantin D. Frank and Sons
  Vinifera Wine Cellars, Ltd.
Vinifera Wine Growers
  Association, 158
Vinterra Farm Winery and
  Vineyard, 103
Von Stiehl Wine, Inc., 191
Vose Vineyards, 179

Wagner Vineyards, 185
Warner Vineyards, 115
Warren J. Sublette Winery,
  102–103
Waukon Corporation, The, 182
Weibel Champagne Vineyards,
  86–87
Wente Brothers, 49
White Mountain Vineyards, Inc.,
  137
Wickford Vineyards, 136
Wickliffe Winery, 187

INDEX OF VINEYARDS

Widmer's Wine Cellars, Inc., 132
Wiederkehr Wine Cellars, Inc., 169
Willoughby Winery, 121
Willow Creek Vineyard, 87
Wilmont Winery, 149–50
Windmill Farms, 185
Wine Cellars, Inc., 159
Winemasters Winery, 50
Wine Museum of San Francisco, The, 50
Wine and the People, 49–50
Winery Rushing, The, 183
Wine Service Cooperative Incorporated, 179
Wittwer Winery, 179
Woburn Winery, 190
Wollersheim, 110
Woodbridge Vineyard Association, 179

Wooden Valley Winery, 51
Woodside Vineyards, 51
Wyandotte Wine Cellar, Inc., 103–104

Yankee Hill Winery, 51
Yerba Buena Industries, 179
York Mountain Winery, 38
Yverdon Vineyards, 68

Zaca Mesa Winery, 179
ZD Winery, 87
ZD Wines, 179
Ziegler Winery, 168
Ziem Vineyards, 182

For a complete list of books available from Penguin in the United States, write to Dept. DG, Penguin Books, 299 Murray Hill Parkway, East Rutherford, New Jersey 07073.

For a complete list of books available from Penguin in Canada, write to Penguin Books Canada Limited, 2801 John Street, Markham, Ontario L3R 1B4.

If you live in the British Isles, write to Dept. EP, Penguin Books Ltd, Harmondsworth, Middlesex.

THE WHOLE-WORLD WINE CATALOG

William I. Kaufman

"When a man drinks wine at dinner," said Plato, "he begins to be better pleased with himself." But you won't be pleased with yourself if you've chosen the wrong wine. Recognizing this, *The Whole-World Wine Catalog* tells you everything you need to know to choose the wines that are right for you—whatever your taste and your budget. Peripatetic gastronome William I. Kaufman wants wine-drinking to be as joyful an experience for you as it has always been for him. To this end, he presents some 2,500 labels for well-known and popular wines from all the wine-producing countries of the world, from the most obvious (France) to the most unlikely (Russia). Each label is explained in detail, and the wine behind it fully described—whether that wine be a "pop" vintage or one of the great classics of the world. Wine labels contain much more information than most people suspect, Mr. Kaufman reminds us—information that wine-shoppers could put to good use if only they understood "these small, informative, and often very beautiful and artistic bearers of facts, the interpretation of which can mean the difference between your pleasure or your disappointment in the wines you have chosen to buy.... Use *The Whole-World Wine Catalog* as your wine cellar," writes Mr. Kaufman. "The catalog was created to help all those who walk into a supermarket or a wine shop or dine in a restaurant, be it humble or deluxe. Knowing what is behind the label—cost, taste, opinion of the producer, and interesting facts about the grapes, winery, and region, plus historical facts about the producer—will help you in your selection and enjoyment of each wine."